The Universe
and Dr. Einstein

LINCOLN BARNETT

The Universe
and Dr. Einstein

With a Foreword by
ALBERT EINSTEIN

WILLIAM SLOANE ASSOCIATES
Publishers   *New York*

To

J. K. B. AND L. H. B.

ACKNOWLEDGMENT

FOR THEIR HELP AND ADVICE IN THE PREPARA-tion of this book I wish to thank Dr. Allen G. Shenstone of the Department of Physics, Princeton University; Dr. Hermann Weyl and Dr. Valentin Bargmann of the Institute for Advanced Study, Princeton, N. J.; and Dr. H. P. Robertson of the California Institute of Technology.

I also wish to express my appreciation to Dr. Harlow Shapley of the Harvard Observatory, who read the original manuscript and made valuable suggestions and criticisms, with particular reference to the sections dealing with astronomy.

I owe my very special gratitude to Dr. William W. Havens, Jr., of the Department of Physics, Columbia University, who read and checked both the original manuscript, prior to its publication in *Harper's* magazine, and the expanded version that appears here; and who patiently and cordially contributed his time and knowledge to the solution of many difficulties presented by the exposition of this material.

FOREWORD BY ALBERT EINSTEIN

ANYONE WHO HAS EVER TRIED TO PRESENT A rather abstract scientific subject in a popular manner knows the great difficulties of such an attempt. Either he succeeds in being intelligible by concealing the core of the problem and by offering to the reader only superficial aspects or vague allusions, thus deceiving the reader by arousing in him the deceptive illusion of comprehension; or else he gives an expert account of the problem, but in such a fashion that the untrained reader is unable to follow the exposition and becomes discouraged from reading any further.

If these two categories are omitted from today's popular scientific literature, surprisingly little remains. But the little that is left is very valuable indeed. It is of great importance that the general public be given an opportunity to experience—consciously and intelligently—the efforts and results of scientific research. It is not sufficient that each result be taken up, elaborated, and applied by a few specialists in the field. Restricting the body of knowledge to a small group deadens the philosophical spirit of a people and leads to spiritual poverty.

I

Foreword

Lincoln Barnett's book represents a valuable contribution to popular scientific writing. The main ideas of the theory of relativity are extremely well presented. Moreover, the present state of our knowledge in physics is aptly characterized. The author shows how the growth of our factual knowledge, together with the striving for a unified theoretical conception comprising all empirical data, has led to the present situation which is characterized—notwithstanding all successes—by an uncertainty concerning the choice of the basic theoretical concepts.

Princeton, New Jersey
September 10, 1948

CARVED IN THE WHITE WALLS OF THE RIVERSIDE
Church in New York, the figures of six hundred great men of the ages—saints, philosophers, kings—stand in limestone immortality, survey-
ing space and time with blank imperishable eyes. One
panel enshrines the geniuses of science, fourteen of
them, spanning the centuries from Hippocrates, who
died around 370 B.C., to Albert Einstein, who was sixty-
nine years old last March. It is noteworthy that Einstein is the only living man in this whole sculptured
gallery of the illustrious dead.

It is equally noteworthy that of the thousands of
people who worship weekly at Manhattan's most spec-
tacular Protestant church, probably 99 per cent would
be hard pressed to explain why Einstein's image is
there. It is there because a generation ago, when the
iconography of the church was being planned, Dr.
Harry Emerson Fosdick wrote letters to a group of
the nation's leading scientists asking them to submit
lists of the fourteen greatest names in scientific history.
Their ballots varied. Most of them included Archime-
des, Euclid, Galileo, and Newton. But on every list
appeared the name of Albert Einstein.

The Universe and Dr. Einstein

The vast gap that has persisted for more than forty years—since 1905, when the Theory of Special Relativity was first published—between Einstein's scientific eminence and public understanding of it is the measure of a gap in American education. Today most newspaper readers know vaguely that Einstein had something to do with the atomic bomb; beyond that his name is simply a synonym for the abstruse. While his theories form part of the body of modern science, they are not yet part of the modern curriculum. It is not surprising therefore that many a college graduate still thinks of Einstein as a kind of mathematical surrealist rather than as the discoverer of certain cosmic laws of immense importance in man's slow struggle to understand physical reality. He does not know that Relativity, over and above its scientific import, comprises a major philosophical system which augments and illumines the reflections of the great epistemologists—Locke, Berkeley, and Hume. Consequently he has very little notion of the vast, arcane, and mysteriously ordered universe in which he dwells.

*   *   *

Dr. Einstein, now professor emeritus at the Institute for Advanced Study at Princeton, is currently hard at work on a problem which has stumped him for more than a quarter-century and which he intends to solve before he dies. His ambition is to complete his "unified field theory," setting forth in one series of equations the laws governing the two fundamental forces of the universe, gravitation and electromagnetism. The significance of this task can be appreciated only when one realizes that virtually all the phenomena of nature are produced by these two primordial forces. Until a hundred years ago electricity and magnetism —while known and studied since early Greek times— were regarded as separate quantities. But the experiments of Oersted and Faraday in the nineteenth century showed that a current of electricity is always surrounded by a magnetic field, and conversely that under certain conditions magnetic forces can induce electrical currents. From these experiments came the discovery of the electromagnetic field through which light waves, radio waves, and all other electromagnetic disturbances are propagated in space. Thus electricity and magnetism may be considered as a single force. Save for gravitation, nearly all other forces in the material universe—frictional forces, chemical forces which hold atoms together in molecules, cohesive forces which bind larger particles of matter, elastic forces which cause bodies to maintain their shape—are of electromagnetic origin; for all of these involve the interplay of matter, and all matter is composed of atoms which in turn are composed of electrical particles. Yet the similarities between gravitational and electromagnetic phenomena are very striking. The planets spin in the gravitational field of the sun; electrons swirl in the electromagnetic field of the atomic nucleus. The earth, moreover, is a big magnet—a peculiar fact which is apparent to anyone who has ever used a compass. The sun is also a magnet. And so are all the stars.

Although many attempts have been made to identify gravitational attraction as an electromagnetic effect, all have failed. Einstein himself thought he had succeeded in 1929 and published a unified field theory which he later scrapped as inadequate. His present objectives are more ambitious, for he is endeavoring now to formulate a set of universal laws that will encompass both the boundless gravitational and electromagnetic fields of interstellar space and the tiny, terrible field inside the atom. In this vast cosmic picture the abyss between macrocosmos and microcosmos— the very big and the very little—will be bridged, and the whole complex of the universe will resolve into a homogeneous fabric in which matter and energy are indistinguishable and all forms of motion from the slow wheeling of the galaxies to the wild flight of electrons become simply changes in the structure and concentration of the primordial field.

Since the aim of science is to describe and explain the world we live in, Einstein would, by thus defining the manifold of nature within the terms of a single harmonious theory, attain its loftiest goal. The meaning of the word "explain," however, suffers a contraction with man's every step in quest of reality. Science cannot yet really "explain" electricity, magnetism, and gravitation; their effects can be measured and predicted, but of their ultimate nature no more is known to the modern scientist than to Thales of Miletus, who

6

first speculated on the electrification of amber around 585 B.C. Most contemporary physicists reject the notion that man can ever discover what these mysterious forces "really" are. Electricity, Bertrand Russell says, "is not a thing, like St. Paul's Cathedral; it is a way in which things behave. When we have told how things behave when they are electrified, and under what circumstances they are electrified, we have told all there is to tell." Until recently scientists would have scorned such a thesis. Aristotle, whose natural science dominated Western thought for two thousand years, believed that man could arrive at an understanding of ultimate reality by reasoning from *self-evident principles*. It is, for example, a self-evident principle that everything in the universe has its proper place, hence one can deduce that objects fall to the ground because that's where they belong, and smoke goes up because that's where *it* belongs. The goal of Aristotelian science was to explain *why* things happen. Modern science was born when Galileo began trying to explain *how* things happen and thus originated the method of controlled experiment which now forms the basis of scientific investigation.

Out of Galileo's discoveries and those of Newton in the next generation there evolved a mechanical universe of forces, pressures, tensions, oscillations, and waves. There seemed to be no process of nature which could not be described in terms of ordinary experience, illustrated by a concrete model or predicted by Newton's amazingly accurate laws of mechanics. But before the turn of the past century certain deviations from these laws became apparent; and though these deviations were slight, they were of such a fundamental nature that the whole edifice of Newton's machine-like universe began to topple. The certainty that science can explain "how" things happen began to dim about twenty years ago. And right now it is a question whether scientific man is in touch with reality at all—or can ever hope to be.

2

THE FACTORS THAT FIRST LED PHYSICISTS TO distrust their faith in a smoothly functioning mechanical universe loomed on the inner and outer horizons of knowledge—in the unseen realm of the atom and in the fathomless depths of intergalactic space. To describe these phenomena quantitatively, two great theoretical systems were developed between 1900 and 1927. One was the Quantum Theory, dealing with the fundamental units of matter and energy. The other was Relativity, dealing with space, time, and the structure of the universe as a whole.

Both are now accepted pillars of modern physical thought. Both describe phenomena in their fields in terms of consistent, mathematical relationships. They do not answer the Newtonian "how" any more than Newton's laws answered the Aristotelian "why." They provide equations, for example, that define with great accuracy the laws governing the radiation and propagation of light. But the actual mechanism by which the atom radiates light and by which light is propagated through space remains one of nature's supreme mysteries. Similarly the laws governing the phenomenon of radioactivity enable scientists to predict that in a given quantity of uranium a certain number of atoms will disintegrate in a certain length of time. But just which atoms will decay and how they are selected for doom are questions that man cannot yet answer.

In accepting a mathematical description of nature, physicists have been forced to abandon the ordinary world of our experience, the world of sense perceptions. To understand the significance of this retreat it is necessary to step across the thin line that divides physics from metaphysics. Questions involving the relationship between observer and reality, subject and object, have haunted philosophical thinkers since the dawn of reason. Twenty-three centuries ago the Greek philosopher Democritus wrote: "Sweet and bitter, cold and warm as well as all the colors, all these things exist but in opinion and not in reality; what really exists are unchangeable particles, atoms, and their motions in empty space." Galileo also was aware of the purely subjective character of sense qualities like color, taste, smell, and sound and pointed out that "they can no more be ascribed to the external objects than can the tickling or the pain caused sometimes by touching such objects."

The English philosopher John Locke tried to penetrate to the "real essence of substances" by drawing a distinction between what he termed the primary and secondary qualities of matter. Thus he considered that shape, motion, solidity, and all geometrical properties were real or primary qualities, inherent in the object itself; while secondary qualities, like colors, sounds, tastes, were simply projections upon the organs of sense. The artificiality of this distinction was obvious to later thinkers.

"I am able to prove," wrote the great German mathematician, Leibnitz, "that not only light, color, heat, and the like, but motion, shape, and extension too are mere apparent qualities." Just as our visual sense, for example, tells us that a golf ball is white, so vision abetted by our sense of touch tells us that it is also round, smooth, and small—qualities that have no more reality, independent of our senses, than the quality which we define by convention as white.

Thus gradually philosophers and scientists arrived at the startling conclusion that since every object is simply the sum of its qualities, and since qualities exist only in the mind, the whole objective universe of matter and energy, atoms and stars, does not exist except as a construction of the consciousness, an edifice of conventional symbols shaped by the senses of man. As Berkeley, the archenemy of materialism, phrased it: "All the choir of heaven and furniture of earth, in a word all those bodies which compose the mighty frame of the world, have not any substance without the mind. . . . So long as they are not actually perceived by me, or do not exist in my mind, or that of any other created spirit, they must either have no existence at all, or else subsist in the mind of some Eternal Spirit." Einstein carried this train of logic to its ultimate limits by showing that even space and time are forms of intuition, which can no more be divorced from consciousness than can our concepts of color, shape, or size. Space has no objective reality except as an order or arrangement of the objects we perceive in it, and time has no independent existence apart from the order of events by which we measure it.

* * *

These philosophical subtleties have a profound bearing on modern science. For with the philosophers' reduction of all objective reality to a shadow-world of perceptions, scientists became aware of the alarming limitations of man's senses. Anyone who has ever thrust a glass prism into a sunbeam and seen the rainbow colors of the solar spectrum refracted on a screen has looked upon the whole range of visible light. For the human eye is sensitive only to the narrow band of radiation that falls between the red and the violet. A difference of a few one hundred thousandths of a centimeter in wave length makes the difference between visibility and invisibility. The wave length of red light is .00007 cm. and that of violet light .00004 cm.

But the sun also emits other kinds of radiation. Infrared rays, for example, with a wave length of .00008 to .032 cm. are just a little too long to excite the retina to an impression of light, though the skin detects their impact as heat. Similarly ultraviolet rays with a wave length of .00003 to .000001 cm. are too short for the eye to perceive but can be recorded on a photographic plate. Photographs can also be made by the "light" of

X-rays which are even shorter than ultraviolet rays. And there are other electromagnetic waves of lesser and greater frequency—the gamma rays of radium, radio waves, cosmic rays—which can be detected in various ways and differ from light only in wavelength. It is evident, therefore, that the human eye suppresses most of the "lights" in the world, and that what man can perceive of the reality around him is distorted and enfeebled by the limitations of his organ of vision. The world would appear far different to him if his eye were sensitive, for example, to X-rays.

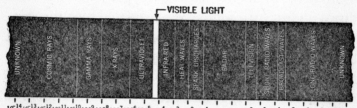

WAVELENGTH (CENTIMETERS)

The electromagnetic spectrum reveals the narrow range of radiation visible to man's eye. From the standpoint of physics, the only difference between radio waves, visible light, and such high-frequency forms of radiation as X-rays and gamma rays lies in their wave length. But out of this vast range of electromagnetic radiation, extending from cosmic rays with wave lengths of only one trillionth of a centimeter up to infinitely long radio waves, the human eye selects only the narrow band indicated in white on the above chart. Man's perceptions of the universe in which he dwells are thus restricted by the limitations of his visual sense. Wave lengths are indicated on the chart by the denary system: i.e. 10^3 centimeters equals 10 x 10 x 10 equals 1,000; and 10^{-3} equals 1/10 x 1/10 x 1/10 equals 1/1,000.

Realization that our whole knowledge of the universe is simply a residue of impressions clouded by our imperfect senses makes the quest for reality seem hopeless. If nothing has existence save in its being perceived, the world should dissolve into an anarchy of individual perceptions. But a curious order runs through our perceptions, as if indeed there might be an underlayer of objective reality which our senses translate. Although no man can ever know whether his sensation of red or of Middle C is the same as another man's, it is nevertheless possible to act on the assumption that everyone sees colors and hears tones more or less alike.

This functional harmony of nature Berkeley, Descartes, and Spinoza attributed to God. Modern physicists who prefer to solve their problems without recourse to God (although this seems to become more difficult all the time) emphasize that nature mysteriously operates on mathematical principles. It is the mathematical orthodoxy of the universe that enables theorists like Einstein to predict and discover natural laws simply by the solution of equations. But the paradox of physics today is that with every improvement in its mathematical apparatus the gulf between man the observer and the objective world of scientific description becomes more profound.

It is perhaps significant that in terms of simple magnitude man is the mean between macrocosm and microcosm. Stated crudely this means that a supergiant red star (the largest material body in the universe) is

The Universe and Dr. Einstein

just as much bigger than man as an electron (tiniest of physical entities) is smaller. It is not surprising, therefore, that the prime mysteries of nature dwell in those realms farthest removed from sense-imprisoned man, nor that science, unable to describe the extremes of reality in the homely metaphors of classical physics should content itself with noting such mathematical relationships as may be revealed.

3

THE FIRST STEP IN SCIENCE'S RETREAT FROM mechanical explanation toward mathematical abstraction was taken in 1900, when Max Planck put forth his Quantum Theory to meet certain problems that had arisen in studies of radiation. It is common knowledge that when heated bodies become incandescent they emit a red glow that turns to orange, then yellow, then white as the temperature increases. Painstaking efforts were made during the past century to formulate a law stating how the amount of radiant energy given off by such heated bodies varied with wave length and temperature. All attempts failed until Planck found by mathematical means an equation that satisfied the results of experiment. The extraordinary feature of his equation was that it rested on the assumption that radiant energy is emitted not in an unbroken stream but in discontinuous bits or portions which he termed *quanta*.

Planck had no evidence for such an assumption, for no one knew anything (then or now) of the actual mechanism of radiation. But on purely theoretical grounds he concluded that each quantum carries an amount of energy given by the equation, $E = h\nu$, where

ν is the frequency of the radiation and h is Planck's Constant, a small but inexorable number (roughly .00000000000000000000000000006624) which has since proved to be one of the most fundamental constants in nature. In any process of radiation the amount of emitted energy divided by the frequency is always equal to h. Although Planck's Constant has dominated the computations of atomic physics for half a century, its magnitude cannot be explained any more than the magnitude of the speed of light can be explained. Like other universal constants it is simply a mathematical fact for which no explanation can be given. Sir Arthur Eddington once observed that any true law of nature is likely to seem irrational to rational man; hence Planck's quantum principle, he thought, is one of the few real natural laws science has revealed.

*　　　*　　　*

The far-reaching implications of Planck's conjecture did not become apparent till 1905, when Einstein, who almost alone among contemporary physicists appreciated its significance, carried the Quantum Theory into a new domain. Planck had believed he was simply patching up the equations of radiation. But Einstein postulated that all forms of radiant energy—light, heat, X-rays—actually travel through space in separate and discontinuous quanta. Thus the sensation of warmth we experience when sitting in front of a fire results from the bombardment of our skin by innumerable quanta of radiant heat. Similarly sensations of color arise from the bombardment of our optic nerves by light quanta which differ from each other just as the frequency ν varies in the equation $E = h\nu$

Einstein substantiated this idea by working out a law accurately defining a puzzling phenomenon known as the photoelectric effect. Physicists had been at a loss to explain the fact that when a beam of pure violet light is allowed to shine upon a metal plate the plate ejects a shower of electrons. If light of lower frequency, say yellow or red, falls on the plate, electrons will again be ejected but at reduced velocities. The vehemence with which the electrons are torn from

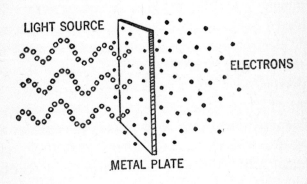

The photoelectric effect was interpreted by Einstein in 1905. When light falls on a metal plate, the plate ejects a shower of electrons. This phenomenon cannot be explained by the classic wave theory of light. Einstein deduced that light is not a continuous stream of energy but is composed of individual particles or bundles of energy which he called *photons*. When a photon strikes an electron the resulting action is analogous to the impact of billiard balls, as shown in the simplified conception above.

the metal depends only on the color of the light and not at all on its intensity. If the light source is removed to a considerable distance and dimmed to a faint glow the electrons that pop forth are fewer in number but their velocity is undiminished. The action is instantaneous even when the light fades to imperceptibility.

Einstein decided that these peculiar effects could be explained only by supposing that all light is composed of individual particles or grains of energy which he called *photons*, and that when one of them hits an electron the resulting action is comparable to the impact of two billiard balls. He reasoned further that photons of violet, ultraviolet, and other forms of high frequency radiation pack more energy than red and infrared photons, and that the velocity with which each electron flies from the metal plate is proportional to the energy content of the photon that strikes it. He expressed these principles in a series of historic equations which won him the Nobel Prize and profoundly influenced later work in quantum physics and spectroscopy. Television and other applications of the photoelectric cell owe their existence to Einstein's Photoelectric Law.

*   *   *

In thus adducing an important new physical principle Einstein uncovered at the same time one of the deepest and most troubling enigmas of nature. No one doubts today that all matter is made up of atoms which in turn are composed of even smaller building blocks called electrons, neutrons, and protons. But Einstein's notion that light too may consist of discontinuous particles clashed with a far more venerable theory that light is made up of waves.

There are indeed certain phenomena involving light that can only be explained by the wave theory. For example the shadows of ordinary objects like buildings, trees and telegraph poles appear sharply defined; but when a very fine wire or hair is held between a light source and a screen it casts no distinct shadow whatsoever, suggesting that light rays have bent around it just as waves of water bend around a small rock. Similarly a beam of light passing through a round aperture projects a sharply-defined disk upon a screen; but if the aperture is reduced to the size of a pinhole, then the disk becomes ribbed with alternating concentric bands of light and darkness, somewhat like those of a conventional target. This phenomenon is known as diffraction and has been compared with the tendency of ocean waves to bend and diverge on passing through the narrow mouth of a harbor. If instead of one pinhole, two pinholes are employed very close together and side by side, the diffraction patterns merge in a series of parallel stripes. Just as two wave systems meeting in a swimming pool will reinforce each other when crest coincides with crest and annul each other when the crest of one wave meets the trough of another, so in the case of the adjacent pinholes the bright stripes occur where two light waves reinforce each other and the dark stripes where two waves have interfered. These phenomena—diffraction and interference—are strictly wave characteristics and would not occur if light were made up of individual corpuscles. More than two centuries of experiment and theory assert that light *must* consist of waves. Yet Einstein's Photoelectric Law shows that light *must* consist of photons.

This fundamental question—is light waves or is it particles?—has never been answered. The dual character of light is, however, only one aspect of a deeper and more remarkable duality which pervades all nature.

* * *

The first hint of this strange dualism came in 1925, when a young French physicist named Louis de Broglie suggested that phenomena involving the interplay of matter and radiation could best be understood by regarding electrons not as individual particles but as systems of waves. This audacious concept flouted two decades of quantum research in which physicists had built up rather specific ideas about the elementary particles of matter. The atom had come to be pictured as a kind of miniature solar system composed of a central nucleus surrounded by varying numbers of electrons (1 for hydrogen, 92 for uranium) revolving in circular or elliptical orbits. The electron was less vivid. Experiments showed that all electrons had exactly the same mass and the same electrical charge, so it was natural to regard them as the ultimate foundation stones of the universe. It also seemed logical at first to picture them simply as hard elastic spheres. But little by little, as investigation progressed, they became more capricious, defiant of observation and measurement. In many ways their behavior appeared too complex for any material particle. "The hard sphere," declared the British physicist, Sir James Jeans, "has always a definite position in space; the electron apparently has not. A hard sphere takes up a very definite amount of room; an electron—well it is probably as meaningless to discuss how much room an electron takes up as it is to discuss how much room a fear, an anxiety, or an uncertainty takes up."

Shortly after de Broglie had his vision of "matter waves" a Viennese physicist named Schrödinger developed the same idea in coherent mathematical form, evolving a system that explained quantum phenomena by attributing specific wave functions to protons and electrons. This system, known as "wave mechanics," was corroborated in 1927 when two American scientists, Davisson and Germer, proved by experiment that electrons actually do exhibit wave characteristics. They directed a beam of electrons upon a metal crystal and obtained diffraction patterns analogous to those produced when light is passed through a pinhole.* Their measurements indicated, moreover, that the wave length of an electron is of the precise magnitude predicted by de Broglie's equation, $\lambda = h/mv$,

---

* A crystal, because of the even and orderly arrangement of its component atoms and the closeness of their spacing, serves as a diffraction grating for very short wave lengths, such as those of X-rays.

where v is the velocity of the electron, m is its mass, and h is Planck's Constant. But further surprises were in store. For subsequent experiments showed that not only electrons but whole atoms and even molecules produce wave patterns when diffracted by a crystal surface, and that their wave lengths are exactly what de Broglie and Schrödinger forecast. And so all the basic units of matter—what J. Clerk Maxwell called "the imperishable foundation stones of the universe" —gradually shed their substance. The old-fashioned spherical electron was reduced to an undulating charge of electrical energy, the atom to a system of superimposed waves. One could only conclude that all matter is made of waves and we live in a world of waves.

The paradox presented by waves of matter on the one hand and particles of light on the other was resolved by several developments in the decade before World War II. The German physicists, Heisenberg and Born, bridged the gap by developing a new mathematical apparatus that permitted accurate description of quantum phenomena either in terms of waves *or* in terms of particles as one wished. The idea behind their system had a profound influence on the philosophy of science. They maintained it is pointless for a physicist to worry about the properties of a single electron; in the laboratory he works with beams or showers of electrons, each containing billions of individual particles (or waves); he is concerned therefore only with mass behavior, with statistics and the laws of probability and chance. So it makes no practical difference whether individual electrons are particles or systems of waves—in aggregate they can be pictured either way. For example, if two physicists are at the seashore one may analyze an ocean wave by saying, "Its properties and intensity are clearly indicated by the positions of its crest and its trough"; while the other may observe with equal accuracy, "The section which you term a crest is significant simply because it contains more molecules of water than the area you call a trough." Analogously Born took the mathematical expression used by Schrödinger in his equations to denote wave function and interpreted it as a "probability" in a statistical sense. That is to say he regarded the intensity of any part of a wave as a measure of the probable distribution of particles at that point. Thus he dealt with the phenomena of diffraction, which hitherto only the wave theory could explain, in terms of the *probability* of certain corpuscles—light quanta or electrons—following certain paths and arriving at certain places. And so "waves of matter" were reduced to "waves of probability." It no longer matters how we visualize an electron or an atom or a probability wave. The equations of Heisenberg and Born fit any picture. And we can, if we choose, imagine ourselves living in a universe of waves, a universe of particles, or as one facetious scientist has phrased it, a universe of "wavicles."

4

WHILE QUANTUM PHYSICS THUS DEFINES with great accuracy the mathematical relationships governing the basic units of radiation and matter, it further obscures the true nature of both. Most modern physicists, however, consider it rather naïve to speculate about the true nature of anything. They are "positivists"—or "logical empiricists" —who contend that a scientist can do no more than report his observations. And so if he performs two experiments with different instruments and one seems to reveal that light is made up of particles and the other that light is made up of waves, he must accept both results, regarding them not as contradictory but as complementary. By itself neither concept suffices to explain light, but together they do. Both are necessary to describe reality and it is meaningless to ask which is really true. For in the abstract lexicon of quantum physics there is no such word as "really."

It is futile, moreover, to hope that the invention of more delicate tools may enable man to penetrate much farther into the microcosm. There is an indeterminacy about all the events of the atomic universe which refinements of measurement and observation can never dispel. The element of caprice in atomic behavior cannot be blamed on man's coarse-grained implements. It stems from the very nature of things, as shown by Heisenberg in 1927 in a famous statement of physical law known as the "Principle of Uncertainty." To illustrate his thesis Heisenberg pictured an imaginary experiment in which a physicist attempts to observe the position and velocity * of a moving electron by using an immensely powerful supermicroscope. Now, as has already been suggested, an individual electron appears to have no definite position or velocity. A physicist can define electron behavior accurately enough so long as he is dealing with great numbers of them. But when he tries to locate a particular electron in space the best he can say is that a certain point in the complex superimposed wave motions of the electron group represents the *probable* position of the electron in question. The individual electron is a blur—as indeterminate as the wind or a sound wave in the night—and the fewer the electrons with which the physicist deals, the more indeterminate his findings. To prove that this indeterminacy is a symptom not of man's immature science but of an ultimate barrier of nature, Heisenberg presupposed that the imaginary microscope used by his imaginary physicist is optically capable of magnifying by a hundred billion diameters—i.e., enough to bring an object the size of an electron within the range of human visibility. But now a further difficulty is encountered. For inasmuch as an electron

* In physics the term "velocity" connotes direction as well as speed.

is smaller than a light wave, the physicist can "illuminate" his subject only by using radiation of shorter wave length. Even X-rays are useless. The electron can be rendered visible only by the high-frequency gamma rays of radium. But the photoelectric effect, it will be recalled, showed that photons of ordinary light exert a violent force on electrons; and X-rays knock them about even more roughly. Hence the impact of a still more potent gamma ray would prove disastrous.

The Principle of Uncertainty asserts therefore that it is absolutely and forever impossible to determine the position and the velocity of an electron at the same time—to state confidently that an electron is "right here at this spot" and is moving at "such and such a speed." For by the very act of observing its position, its velocity is changed; and, conversely, the more accurately its velocity is determined, the more indefinite its position becomes. And when the physicist computes the mathematical margin of uncertainty in his measurements of an electron's position and velocity he finds it is always a function of that mysterious quantity —Planck's Constant, h.

*   *   *

Quantum physics thus demolishes two pillars of the old science, causality and determinism. For by dealing in terms of statistics and probabilities it abandons all idea that nature exhibits an inexorable sequence of cause and effect. And by its admission of margins of uncertainty it yields up the ancient hope that science, given the present state and velocity of every material body in the universe, can forecast the history of the universe for all time. One by-product of this surrender is a new argument for the existence of free will. For if physical events are indeterminate and the future is unpredictable, then perhaps the unknown quantity called "mind" may yet guide man's destiny among the infinite uncertainties of a capricious universe. But this notion invades a realm of thought with which the physicist is not concerned. Another conclusion of greater scientific importance is that in the evolution of quantum physics the barrier between man, peering dimly through the clouded windows of his senses, and whatever objective reality may exist has been rendered almost impassable. For whenever he attempts to penetrate and spy on the "real" objective world, he changes and distorts its workings by the very process of his observation. And when he tries to divorce this "real" world from his sense perceptions he is left with nothing but a mathematical scheme. He is indeed somewhat in the position of a blind man trying to discern the shape and texture of a snowflake. As soon as it touches his fingers or his tongue it dissolves. A wave electron, a photon, a wave of probability, cannot be visualized; they are simply symbols useful in expressing the mathematical relationships of the microcosm.

To the question, why does modern physics employ such abstract methods of description, the physicist answers: because the equations of quantum physics define more accurately than any mechanical model the fundamental phenomena beyond the range of vision. In short, *they work,* as the calculations which hatched the atomic bomb spectacularly proved. The aim of the practical physicist, therefore, is to enunciate the laws of nature in ever more precise mathematical terms. Where the nineteenth century physicist envisaged electricity as a fluid and, with this metaphor in mind, evolved the laws that generated our present electrical age, the twentieth century physicist tends to avoid metaphors. He knows that electricity is not a fluid, and he knows that such pictorial concepts as "waves" and "particles," while serving as guideposts to new discovery, must not be accepted as accurate representations of reality. In the abstract language of mathematics he can describe how things behave though he does not know—or need to know—what they are.

Yet there are present-day physicists for whom the void between science and reality presents a challenge. Einstein has more than once expressed the hope that the statistical method of quantum physics would prove a temporary expedient. "I cannot believe," he says, "that God plays dice with the world." He repudiates the positivist doctrine that science can only report and correlate the results of observation. He believes in a universe of order and harmony. And he believes that questing man may yet attain a knowledge of ultimate reality. To this end he has looked not within the atom, but outward to the stars, and beyond them to the vast drowned depths of empty space and time.

5

IN HIS GREAT TREATISE *On Human Understanding* philosopher John Locke wrote three hundred years ago: "A company of chessmen standing on the same squares of the chessboard where we left them, we say, are all in the same place or unmoved: though perhaps the chessboard has been in the meantime carried out of one room into another. . . . The chessboard, we also say, is in the same place if it remain in the same part of the cabin, though perhaps the ship which it is in sails all the while; and the ship is said to be in the same place supposing it kept the same distance with the neighboring land, though perhaps the earth has turned around; and so chessmen and board and ship have every one changed place in respect to remoter bodies."

Embodied in this little picture of the moving but unmoved chessmen is one principle of relativity—relativity of position. But this suggests another idea—relativity of motion. Anyone who has ever ridden on a railroad train knows how rapidly another train flashes by when it is traveling in the opposite direction, and conversely how it may look almost motionless when it is moving in the same direction. A variation of this effect can be very deceptive in an enclosed station like Grand Central Terminal in New York. Once in a while a train gets under way so gently that passengers feel no recoil whatever. Then if they happen to look out the window and see another train slide past on the next track, they have no way of knowing which train is in motion and which is at rest; nor can they tell how fast either one is moving or in what direction. The only way they can judge their situation is by looking out the other side of the car for some fixed body of reference like the station platform or a signal light. Sir Isaac Newton was aware of these tricks of motion, only he thought in terms of ships. He knew that on a calm day at sea a sailor can shave himself or drink soup as comfortably as when his ship is lying motionless in harbor. The water in his basin, the soup in his bowl, will remain unruffled whether the ship is making five knots, 15 knots, or 25 knots. So unless he peers out at the sea it will be possible for him to know how fast his ship is moving or indeed if it is moving at all. Of course if the sea should get rough or the ship change course abruptly, then he will sense his state of motion. But granted the idealized conditions of a glass-calm sea and a silent ship, nothing that happens below decks —no amount of observation or mechanical experiment performed *inside* the ship—will disclose its velocity through the sea. The physical principle suggested by these considerations was formulated by Newton in 1687. "The motions of bodies included in a given space," he wrote, "are the same among themselves, whether that space is at rest or moves uniformly forward in a straight line." This is known as the Newtonian or Galilean Relativity Principle. It can also be phrased in more general terms: mechanical laws which are valid in one place are equally valid in any other place which moves uniformly relative to the first.

*　　*　　*

The philosophical importance of this principle lies in what it says about the universe. Since the aim of science is to explain the world we live in, as a whole and in all its parts, it is essential to the scientist that he have confidence in the harmony of nature. He must believe that physical laws revealed to him on earth are in truth universal laws. Thus in relating the fall of an apple to the wheeling of the planets around the sun Newton hit upon a universal law. And although he illustrated his principle of relative motion by a ship at sea, the ship he actually had in mind was the earth. For all ordinary purposes of science the earth can be regarded as a stationary system. We may say if we choose that mountains, trees, houses, are at rest, and animals, automobiles, and airplanes move. But to the astrophysicist, the earth, far from being at rest, is whirling through space in a giddy and highly complicated fashion. In addition to its daily rotation about its axis at the rate of 1000 miles an hour, and its annual revolution about the sun at the rate of 20 miles a second, the earth is also involved in a number of other less familiar gyrations. Contrary to popular belief the moon does not revolve around the earth; they revolve around each other—or more precisely, around a common center of gravity. The entire solar system, moreover, is moving within the local star system at the rate of 13 miles a second; the local star system is moving within the Milky Way at the rate of 200 miles a second; and the whole Milky Way is drifting with respect to the remote external galaxies at the rate of 100 miles a second—and all in different directions!

Although he could not then know the full complexity of the earth's movements, Newton was nevertheless troubled by the problem of distinguishing relative motion from true or "absolute" motion in a confusingly busy universe. He suggested that "in the remote regions of the fixed stars or perhaps far beyond them, there may be some body absolutely at rest," but admitted there was no way of proving this by any celestial object within man's view. On the other hand it seemed to Newton that space itself might serve as a fixed frame of reference to which the wheeling of the stars and galaxies could be related in terms of absolute motion. He regarded space as a physical reality, stationary and immovable; and while he could not support this conviction by any scientific argument, he nevertheless clung to it on theological grounds. For to Newton space represented the divine omnipresence of God in nature.

In the next two centuries it appeared probable that Newton's view would prevail. For with the development of the wave theory of light scientists found it necessary to endow empty space with certain mechanical properties—to assume, indeed, that space was some kind of substance. Even before Newton's time the French philosopher, Descartes, had argued that the mere separation of bodies by distance proved the existence of a medium between them. And to eighteenth and nineteenth century physicists it was obvious that if light consisted of waves, there must be some medium to support them, just as water propagates the waves of the sea and air transmits the vibrations we call sound. Hence when experiments showed that light can travel in a vacuum, scientists evolved a hypothetical substance called "ether" which they decided must pervade all space and matter. Later on Faraday propounded another kind of ether as the carrier of electric and magnetic forces. When Maxwell finally identified light as an electromagnetic disturbance the case for the ether seemed assured.

*   *   *

A universe permeated with an invisible medium in which the stars wandered and through which light traveled like vibrations in a bowl of jelly was the end product of Newtonian physics. It provided a mechanical model for all known phenomena of nature, and it provided the fixed frame of reference, the absolute and immovable space, which Newton's cosmology required. Yet the ether presented certain problems, not the least of which was that its actual existence had never been proved. To discover once and for all

The Universe and Dr. Einstein

whether there really was any such thing as ether, two American physicists, A. A. Michelson and E. W. Morley, performed a classic experiment in Cleveland in the year 1881.

The principle underlying their experiment was quite simple. They reasoned that if all space is simply a motionless sea of ether, then the earth's motion through the ether should be detectable and measurable in the same way that sailors measure the velocity of a ship through the sea. As Newton pointed out, it is impossible to detect the movement of a ship through calm waters by any mechanical experiment performed *inside* the ship. Sailors ascertain a ship's speed by throwing a log overboard and watching the unreeling of the knots on the log line. Hence to detect the earth's motion through the ether sea, Michelson and Morley threw a "log" overboard, and the log was a beam of light. For if light really is propagated through the ether, then its velocity should be affected by the ether stream arising from the earth's movement. Specifically a light ray projected in the direction of the earth's movement should be slightly retarded by the ether flow, just as a swimmer is retarded by a current when going upstream. The difference would be slight, for the velocity of light (which was accurately determined in 1849) is 186,284 miles a second, while the velocity of the earth in its orbit around the sun is only 20 miles a second. Hence a light ray sent *against* the ether stream should travel at the rate of 186,264 miles a second, while one sent *with* the ether stream should

35

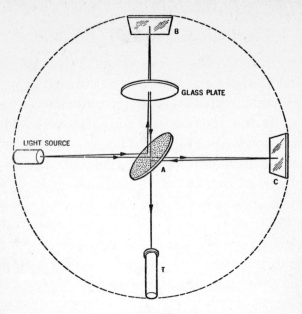

The Michelson-Morley interferometer consisted of an arrangement of mirrors, so designed that a beam transmitted from a light source (above left) was divided and sent in two directions at the same time. This was done by a mirror, A, the *face* of which was only thinly silvered, so that part of the beam was permitted to pass through to mirror C (right) and the remainder reflected at right angles toward mirror B. Mirrors B and C then reflected the rays back to mirror A where, reunited, they proceeded to an observing telescope T. Since the beam ACT had to pass three times through the thickness of glass behind the reflecting face of mirror A, a clear glass plate of equal thickness was placed between A and B to intercept beam ABT and compensate for this retardation. The whole apparatus was rotated in different directions so that the beams ABT and ACT could be sent with, against, and at right angles to the postulated ether stream. At first glance it might appear that a trip "downstream", for example from B to A, should compensate in time for an "upstream" trip from A to B. But this is not so. To row a boat one mile upstream and another mile downstream takes longer than rowing two miles in still water or across current, even with allowance for drift. Had there been any acceleration or retardation of either beam by the ether stream, the optical apparatus at T would have detected it.

be clocked at 186,304 miles a second. With these ideas in mind Michelson and Morley constructed an instrument of such great delicacy that it could detect a variation of even a fraction of a mile per second in the enormous velocity of light. This instrument, which they called an "interferometer" consisted of a group of mirrors so arranged that a light beam could be split in two and flashed in different directions at the same time. The whole experiment was planned and executed with such painstaking precision that the result could not be doubted. And the result was simply this: there was no difference whatsoever in the velocity of the light beams regardless of their direction.

The Michelson-Morley experiment confronted scientists with an embarrassing alternative. On the one hand they could scrap the ether theory which had explained so many things about electricity, magnetism, and light. Or if they insisted on retaining the ether they had to abandon the still more venerable Copernican theory that the earth is in motion. To many physicists it seemed almost easier to believe that the earth stood still than that waves—light waves, electromagnetic waves—could exist without a medium to sustain them. It was a serious dilemma and one that split scientific thought for a quarter century. Many new hypotheses were advanced and rejected. The experiment was tried again by Morley and by others, with the same conclusion; the apparent velocity of the earth through the ether was zero.

6

AMONG THOSE WHO PONDERED THE ENIGMA OF THE Michelson-Morley experiment was a young patent office examiner in Berne, named Albert Einstein. In 1905, when he was just twenty-six years old, he published a short paper suggesting an answer to the riddle in terms that opened up a new world of physical thought. He began by rejecting the ether theory and with it the whole idea of space as a fixed system or framework, absolutely at rest, within which it is possible to distinguish absolute from relative motion. The one indisputable fact established by the Michelson-Morley experiment was that the velocity of light is unaffected by the motion of the earth. Einstein seized on this as a revelation of universal law. If the velocity of light is constant regardless of the earth's motion, he reasoned, it must be constant regardless of the motion of any sun, moon, star, meteor, or other system moving anywhere in the universe. From this he drew a broader generalization, and asserted that the laws of nature are the same for all uniformly moving systems. This simple statement is the essence of Einstein's Special Theory of Relativity. It incorporate the Galilean Relativity Principle which stated that mechanical laws are the same for all uniformly moving systems. But its phrasing is more comprehensive; for Einstein was thinking not only of mechanical laws but of the laws governing light and other electromagnetic phenomena. So he lumped them together in one fundamental postulate: all the phenomena of nature, all the laws of nature, are the same for all systems that move uniformly relative to one another.

On the surface there is nothing very startling in this declaration. It simply reiterates the scientist's faith in the universal harmony of natural law. It also advises the scientist to stop looking for any absolute, stationary frame of reference in the universe. The universe is a restless place: stars, nebulae, galaxies, and all the vast gravitational systems of outer space are incessantly in motion. But their movements can be described only with respect to each other, for in space there are no directions and no boundaries. It is futile moreover for the scientist to try to discover the "true" velocity of any system by using light as a measuring rod, for the velocity of light is constant throughout the universe and is unaffected either by the motion of its source or the motion of the receiver. Nature offers no absolute standards of comparison; and space is—as another great German mathematician, Leibnitz, clearly saw two centuries before Einstein—simply "the order or relation of things among themselves." Without things occupying it, it is nothing.

Along with absolute space, Einstein discarded the concept of absolute time—of a steady, unvarying, inexorable universal time flow, streaming from the infi-

nite past to the infinite future. Much of the obscurity
that has surrounded the Theory of Relativity stems
from man's reluctance to recognize that sense of time,
like sense of color, is a form of perception. Just as there
is no such thing as color without an eye to discern it,
so an instant or an hour or a day is nothing without an
event to mark it. And just as space is simply a possible
order of material objects, so time is simply a possible
order of events. The subjectivity of time is best explained in Einstein's own words. "The experiences of
an individual," he says, "appear to us arranged in a
series of events; in this series the single events which
we remember appear to be ordered according to the
criterion of 'earlier' and 'later.' There exists, therefore,
for the individual, an I-time, or subjective time. This
in itself is not measurable. I can, indeed, associate
numbers with the events, in such a way that a greater
number is associated with the later event than with an
earlier one. This association I can define by means of
a clock by comparing the order of events furnished by
the clock with the order of the given series of events.
We understand by a clock something which provides a
series of events which can be counted."

By referring our own experiences to a clock (or a
calendar) we make time an objective concept. Yet the
time intervals provided by a clock or a calendar are
by no means absolute quantities imposed on the entire
universe by divine edict. All the clocks ever used by
man have been geared to our solar system. What we
call an hour is actually a measurement in space—an

arc of 15 degrees in the apparent daily rotation of the celestial sphere. And what we call a year is simply a measure of the earth's progress in its orbit around the sun. An inhabitant of Mercury, however, would have very different notions of time. For Mercury makes its trip around the sun in 88 of our days, and in that same period rotates just once on its axis. So on Mercury a year and a day amount to the same thing. But it is when science ranges beyond the neighborhood of the sun that all our terrestrial ideas of time become meaningless. For Relativity tells us there is no such thing as a fixed interval of time independent of the system to which it is referred. There is indeed no such thing as simultaneity, there is no such thing as "now," independent of a system of reference. For example a man in New York may telephone a friend in London, and although it is 7:00 P.M. in New York and midnight in London, we may say that they are talking "at the same time." But that is because they are both residents of the same planet, and their clocks are geared to the same astronomical system. A more complicated situation arises if we try to ascertain, for example, what is happening on the star Arcturus "right now." Arcturus is 38 light years away. A light year is the distance light travels in one year, or roughly six trillion miles. If we should try to communicate with Arcturus by radio "right now" it would take 38 years for our message to reach its destination and another 38 years for us to receive a reply.* And when we look at Arcturus and say

* Radio waves travel at the same speed as light waves.

that we see it "now," we are actually seeing a ghost—
an image projected on our optic nerves by light rays
that left their source in 1910. Whether Arcturus even
exists "now" nature forbids us to know until 1986.

*　　　*　　　*

Despite such reflections it is difficult for earthbound
man to accept the idea that *this very instant* which he
calls "now" cannot apply to the universe as a whole.
Yet in the Special Theory of Relativity Einstein
proves by an unanswerable sequence of example and
deduction that it is nonsense to think of events taking
place simultaneously in unrelated systems. His argument unfolds along the following lines.

To begin with one must realize that the scientist,
whose task it is to describe physical events in objective
terms, cannot use subjective words like "this," "here,"
and "now." For him concepts of space and time take
on physical significance only when the relations between events and systems are defined. And it is constantly necessary for him, in dealing with matters involving complex forms of motion (as in celestial
mechanics, electrodynamics, etc.) to relate the magni-
tudes found in one system with those occurring in another. The mathematical laws which define these
relationships are known as laws of transformation.
The simplest transformation may be illustrated by a
man promenading on the deck of a ship: if he walks
forward along the deck at the rate of 3 miles an hour
and the ship moves through the sea at the rate of 12

miles an hour, then the man's velocity with respect to the sea is 15 miles an hour; if he walks aft his velocity relative to the sea is of course 9 miles an hour. Or as a variation one may imagine an alarm bell ringing at a railway crossing. The sound waves produced by the bell spread away through the surrounding air at the rate of 400 yards a second. A railroad train speeds toward the crossing at the rate of 20 yards a second. Hence the velocity of the sound relative to the train is 420 yards a second so long as the train is approaching the alarm bell and 380 yards a second as soon as the train passes the bell. This simple addition of velocities rests on obvious common sense, and has indeed been applied to problems of compound motion since the time of Galileo. Serious difficulties arise, however, when it is used in connection with light.

In his original paper on Relativity Einstein emphasized these difficulties with another railway incident. Again there is a crossing, marked this time by a signal light which flashes its beam down the track at 186,284 miles a second—the constant velocity of light, denoted in physics by the symbol c. A train steams toward the signal light at a given velocity v. So by the addition of velocities one concludes that the velocity of the light beam relative to the train is c plus v when the train moves toward the signal light, and c minus v as soon as the train passes the light. But this result conflicts with the findings of the Michelson-Morley experiment which demonstrated that the velocity of light is unaffected either by the motion of the source or the motion of the receiver. This curious fact has also been confirmed by studies of double stars which revolve around a common center of gravity. Careful analysis of these moving systems has shown that the light from the approaching star in each pair reaches earth at precisely the same velocity as the light from the receding star. Since the velocity of light is a universal constant it cannot in Einstein's railway problem be affected by the velocity of the train. Even if we imagine that the train is racing toward the signal light at a speed of 10,000 miles a second, the principle of the constancy of the velocity of light tells us that an observer aboard the train will still clock the speed of the oncoming light beam at precisely 186,284 miles a second, no more, no less.

The dilemma presented by this situation involves much more than a Sunday morning newspaper puzzle. On the contrary it poses a deep enigma of nature. Einstein saw that the problem lay in the irreconcilable conflict between his belief in (1) the constancy of the velocity of light, and (2) the principle of the addition of velocities. Although the latter appears to rest on the stern logic of mathematics (i.e., that two plus two makes four), Einstein recognized in the former a fundamental law of nature. He concluded, therefore, that a new transformation rule must be found to enable the scientist to describe the relations between moving systems in such a way that the results satisfy the known facts about light.

*   *   *

Einstein found what he wanted in a series of equations developed by the great Dutch physicist, H. A. Lorentz, in connection with a specific theory of his own. Although its original application is of interest now chiefly to scientific historians, the Lorentz transformation lives on as part of the mathematical framework of Relativity. To understand what it says, however, it is first necessary to perceive the flaws in the old principle of the addition of velocities. These flaws Einstein pointed out by means of still another railway anecdote. Once again he envisaged a straight length of track, this time with an observer sitting on an embankment beside it. A thunderstorm breaks, and two bolts of lightning strike the track simultaneously at

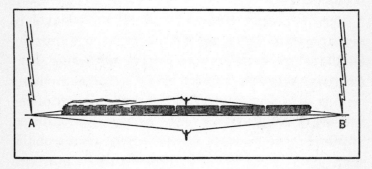

separate points, A and B. Now, asks Einstein, what do we mean by "simultaneously"? To pin down this definition he assumes that the observer is sitting precisely half way between A and B, and that he is equipped with an arrangement of mirrors which enable him to see A and B at the same time without moving his eyes. Then if the lightning flashes are reflected in the ob-

The Universe and Dr. Einstein

server's mirrors at precisely the same instant, the two
flashes may be regarded as simultaneous. Now a train
roars down the track, and a second observer is sitting
precariously perched atop one of the cars with a mirror apparatus just like the one on the embankment. It
happens that this moving observer finds himself directly opposite the observer on the embankment at the
precise instant the lightning bolts hit A and B. The
question is: will the lightning flashes appear simultaneous to him? The answer is: they will not. For if his
train is moving away from lightning bolt B and toward lightning bolt A, then it is obvious that B will be
reflected in his mirrors a fraction of a second later than
A. Lest there be any doubt about this, one may imagine
temporarily that the train is moving at the impossible
rate of 186,284 miles a second, the velocity of light. In
that event flash B will never be reflected in the mirrors
because it can never overtake the train, just as the
sound from a gun can never overtake a bullet traveling
with supersonic speed. So the observer on the train will
assert that only one lightning bolt struck the track.
And whatever the speed of the train may be the moving observer will always insist that the lightning flash
ahead of him has struck the track first. Hence the lightning flashes which are simultaneous relative to the stationary observer are *not* simultaneous relative to the
observer on the train.

The paradox of the lightning flashes thus drama-
tizes one of the subtlest and most difficult concepts in
Einstein's philosophy: the relativity of simultaneity.

The Universe and Dr. Einstein

It shows that man cannot assume that his subjective sense of "now" applies to all parts of the universe. For, Einstein points out, "every reference body (or co-ordinate system) has its own particular time; unless we are told the reference body to which the statement of time refers, there is no meaning in a statement of the time of an event." The fallacy in the old principle of the addition of velocities lies therefore in its tacit assumption that the duration of an event is independent of the state of motion of the system of reference. In the case of the man pacing the deck of a ship, for example, it was assumed that if he walked three miles in one hour as timed by a clock on the moving ship, his rate would be just the same timed by a stationary clock anchored somehow in the sea. It was further assumed that the distance he traversed in one hour would have the same value whether it was measured relative to the deck of the ship (the moving system) or relative to the sea (the stationary system). This constitutes a second fallacy in the addition of velocities—for distance, like time, is a relative concept, and there is no such thing as a space interval independent of the state of motion of the system of reference.

Einstein asserted, therefore, that the scientist who wishes to describe the phenomena of nature in terms that are consistent for all systems throughout the universe must regard measurements of time and distance as variable quantities. The equations comprising the Lorentz transformation do just that. They preserve the velocity of light as a universal constant, but modify all measurements of time and distance according to the velocity of each system of reference.*

So although Lorentz had originally developed his equations to meet a specific problem, Einstein made them the basis of a tremendous generalization, and to the edifice of Relativity added another axiom: the

* The Lorentz transformation relates distances and times observed on moving systems with those observed on systems relatively at rest. Suppose, for example, that a system, or reference body, is moving in a certain direction, then *according to the old principle of the addition of velocities,* a distance or length x', measured with respect to the moving system along the direction of motion, is related to length x, measured with respect to a relatively stationary system, by the equation $x' = x \pm vt$, where v is the velocity of the moving system and t is the time. Dimensions y' and z', measured with respect to the moving system at right angles to x' and at right angles to each other (i.e., height and breadth), are related to dimensions y and z on the relatively stationary system by $y' = y$, and $z' = z$. And finally a time interval t', clocked with respect to the moving system, is related to time interval t, clocked with respect to the relatively stationary system, by $t' = t$. In other words, distances and times are not affected, *in classical physics,* by the velocity of the system in question. *But it is this presupposition which leads to the paradox of the lightning flashes.* The Lorentz transformation reduces the distances and times observed on moving systems to the conditions of the stationary observer, keeping the velocity of light c a constant for all observers. Here are the equations of the Lorentz transformation which have supplanted the older and evidently inadequate relationships cited above:

$$x' = \sqrt{\frac{x - vt}{1 - (v^2/c^2)}}$$
$$y' = y$$
$$z' = z$$
$$t' = \sqrt{\frac{t - (v/c^2)x}{1 - (v^2/c^2)}}$$

It will be noted that, as in the old transformation law, dimensions y' and z' are unaffected by motion. It will also be seen that if the velocity of the moving system v is small relative to the velocity of light c, then the equations of the Lorentz transformation reduce themselves to the relations of the old principle of the addition of velocities. But as the magnitude of v approaches that of c, then the values of x' and t' are radically changed.

laws of nature preserve their uniformity in all systems when related by the Lorentz transformation. Stated thus, in the abstract language of mathematics the significance of this axiom can scarcely be apparent to the layman. But in physics an equation is never a pure abstraction; it is simply a kind of shorthand expression which the scientist finds convenient to describe the phenomena of nature. Sometimes it is also a Rosetta Stone in which the theoretical physicist can decipher secret realms of knowledge. And so by deduction from the message written in the equations of the Lorentz transformation, Einstein discovered a number of new and extraordinary truths about the physical universe.

7

THESE TRUTHS CAN BE DESCRIBED IN VERY CON-crete terms. For once he had evolved the phil-osophical and mathematical bases of Rela-tivity, Einstein had to bring them into the laboratory, where abstractions like time and space are harnessed by means of clocks and measuring rods. And so trans-lating his basic ideas about time and space into the language of the laboratory, he pointed out some hither-to unsuspected properties of clocks and rods. For ex-ample: a clock attached to any moving system runs at a different rhythm from a stationary clock; and a measuring rod attached to any moving system changes its length according to the velocity of the system. Spe-cifically the clock slows down as its velocity increases, and the measuring rod shrinks in the direction of its motion. These peculiar changes have nothing to do with the construction of the clock or the composition of the rod. The clock can be a pendulum clock, a spring clock, or an hour glass. The measuring rod can be a wooden ruler, a metal yardstick, or a ten-mile cable. The slowing of the clock and the contraction of the rod are not mechanical phenomena; an observer rid-ing along with the clock and the measuring rod would not notice these changes. But a stationary observer, i.e., stationary relative to the moving system, would find that the moving clock has slowed down with respect to his stationary clock, and that the moving rod has contracted with respect to his stationary units of measurement.

This singular behavior of moving clocks and yardsticks accounts for the constant velocity of light. It explains why all observers in all systems everywhere, regardless of their state of motion, will always find that light strikes their instruments and departs from their instruments at precisely the same velocity. For as their own velocity approaches that of light, their clocks slow down, their yardsticks contract, and all their measurements are reduced to the values obtained by a relatively stationary observer. The laws governing these contractions are defined by the Lorentz transformation and they are very simple: the greater the speed, the greater the contraction. A yardstick moving with 90 per cent the velocity of light would shrink to about half its length; thereafter the rate of contraction becomes more rapid; and if the stick could attain the velocity of light, it would shrink away to nothing at all. Similarly a clock traveling with the velocity of light would stop completely. From this it follows that nothing can ever move faster than light, no matter what forces are applied. Thus Relativity reveals another fundamental law of nature: *the velocity of light is the top limiting velocity in the universe.*

At first meeting these facts are difficult to digest but

that is simply because classical physics assumed, unjustifiably, that an object preserves the same dimensions whether it is in motion or at rest and that a clock keeps the same rhythm in motion and at rest. Common sense dictates that this must be so. But as Einstein has pointed out, common sense is actually nothing more than a deposit of prejudices laid down in the mind prior to the age of eighteen. Every new idea one encounters in later years must combat this accretion of "self-evident" concepts. And it is because of Einstein's unwillingness ever to accept any unproven principle as self-evident that he was able to penetrate closer to the underlying realities of nature than any scientist before him. Why, he asked, is it any more strange to assume that moving clocks slow down and moving rods contract, than to assume that they don't? The reason classical physics took the latter view for granted is that man, in his everyday experience, never encounters velocities great enough to make these changes manifest. In an automobile, an airplane, even in a V-2 rocket, the slowing down of a watch is immeasurable. It is only when velocities approximate that of light that relativistic effects can be detected. The equations of the Lorentz transformation show very plainly that at ordinary speeds the modification of time and space intervals amounts practically to zero. Relativity does not therefore contradict classical physics. It simply regards the old concepts as limiting cases that apply solely to the familiar experiences of man.

*   *   *

Einstein thus surmounts the barrier reared by man's impulse to define reality solely as he perceives it through the screen of his senses. Just as the Quantum Theory demonstrated that elementary particles of matter do not behave like the larger particles we discern in the coarse-grained world of our perceptions, so Relativity shows that we cannot foretell the phenomena accompanying great velocities from the sluggish behavior of objects visible to man's indolent eye. Nor may we assume that the laws of Relativity deal with exceptional occurrences; on the contrary they provide a comprehensive picture of an incredibly complex universe in which the simple mechanical events of our earthly experience are the exceptions. The present-day scientist, coping with the tremendous velocities that prevail in the fast universe of the atom or with the immensities of sidereal space and time, finds the old Newtonian laws inadequate. But Relativity provides him in every instance with a complete and accurate description of nature.

Whenever Einstein's postulates have been put to test, their validity has been amply confirmed. Remarkable proof of the relativistic retardation of time intervals came out of an experiment performed by H. E. Ives of the Bell Telephone Laboratories in 1936. A radiating atom may be regarded as a kind of clock in that it emits light of a definite frequency and wavelength which can be measured with great precision by means of a spectroscope. Ives compared the light emitted by hydrogen atoms moving at high velocities with that emitted by hydrogen atoms at rest, and found that the frequency of vibration of the moving atoms was reduced in exact accordance with the predictions of Einstein's equations. Someday science may devise a far more interesting test of the same principle. Since any periodic motion serves to measure time, the human heart, Einstein has pointed out, is also a kind of clock. Hence, according to Relativity, the heartbeat of a person traveling with a velocity close to that of light would be relatively slowed, along with his respiration and all other physiological processes. He would not notice this retardation because his watch would slow down in the same degree. But judged by a stationary timekeeper he would "grow old" less rapidly. In a Buck Rogers realm of fantasy, it is possible to imagine some future cosmic explorer boarding an atom-propelled space ship, ranging the void at 167,000 miles per second, and returning to earth aften ten terrestrial years to find himself physically only five years older.

8

IN ORDER TO DESCRIBE THE MECHANICS OF THE physical universe, three quantities are required: time, distance, and mass. Since time and distance are relative quantities one might guess that the mass of a body also varies with its state of motion. And indeed the most important practical results of Relativity have arisen from this principle—the relativity of mass.

In its popular sense, "mass" is just another word for "weight." But as used by the physicist, it denotes a rather different and more fundamental property of matter: namely, resistance to a change of motion. A greater force is necessary to move a freight car than a velocipede; the freight car resists motion more stubbornly than the velocipede because it has greater mass. In classical physics the mass of any body is a fixed and unchanging property. Thus the mass of a freight car should remain the same whether it is at rest on a siding, rolling across country at 60 miles an hour, or hurtling through outer space at 60,000 miles a second. But Relativity asserts that the mass of a moving body is by no means constant, but increases with its velocity. The old physics failed to discover this fact simply because man's senses and instruments are too crude to note the infinitesimal increases of mass produced by the feeble

The Universe and Dr. Einstein

accelerations of ordinary experience. They become perceptible only when bodies attain velocities close to that of light. (This phenomenon, incidentally, does not conflict with the relativistic contraction of length. One is tempted to ask: how can an object become smaller and at the same time get heavier? The contraction, it should be noted, is only in the direction of motion; width and breadth are unaffected. Moreover mass is not "heaviness" but resistance to motion.)

Einstein's equation giving the increase of mass with velocity is similar in form to the other equations of Relativity but vastly more important in its consequences:

$$m = \frac{m_o}{\sqrt{1 - v^2/c^2}}$$

Here m stands for the mass of a body moving with velocity v, m_o for its mass when at rest, and c for the velocity of light. Anyone who has ever studied elementary algebra can readily see that if v is small, as are all the velocities of ordinary experience, then the difference between m_o and m is practically zero. But when v approaches the value of c then the increase of mass becomes very great, reaching infinity when the velocity of the moving body reaches the velocity of light. Since a body of infinite mass would offer infinite resistance to motion the conclusion is once again reached that no material body can travel with the speed of light.*

Of all aspects of Relativity the principle of increase

* See Appendix.

of mass has been most often verified and most fruit-fully applied by experimental physicists. Electrons moving in powerful electrical fields and beta particles ejected from the nuclei of radioactive substances attain velocities ranging up to 99 per cent that of light. For atomic physicists concerned with these great speeds, the increase of mass predicted by Relativity is no argu-able theory but an empirical fact their calculations cannot ignore. In fact the mechanics of the prota-syn-chrotron and other new super-energy machines are de-signed to allow for the increasing mass of particles as their speed approaches the velocity of light.

By further deduction from his principle of Rela-tivity of mass, Einstein arrived at a conclusion of incal-culable importance to the world. His train of reason-ing ran somewhat as follows: since the mass of a mov-ing body increases as its motion increases, and since motion is a form of energy (kinetic energy), then the increased mass of a moving body comes from its in-creased energy. In short, energy has mass! By a few comparatively simple mathematical steps, Einstein found the value of the equivalent mass m in any unit of energy E and expressed it by the equation $m = E/c^2$. Given this relation a high school freshman can take the remaining algebraic step necessary to write the most important and certainly the most famous equa-tion in history: $E = mc^2$.

The part played by this equation in the development of the atomic bomb is familiar to most newspaper readers. It states in the shorthand of physics that the energy contained in any particle of matter is equal to the mass of that body (in grams) multiplied by the square of the velocity of light (in centimeters per second). This extraordinary relationship becomes more vivid when its terms are translated into concrete values: i.e., one kilogram of coal (about two pounds), if converted *entirely* into energy, would yield 25 billion kilowatt hours of electricity or as much as all the power plants in the U.S. could generate by running steadily for two months.

$E = mc^2$ provides the answer to many of the long-standing mysteries of physics. It explains how radioactive substances like radium and uranium are able to eject particles at enormous velocities and to go on doing so for millions of years. It explains how the sun and all the stars can go on radiating light and heat for billions of years; for if our sun were being consumed by ordinary processes of combustion, the earth would have died in frozen darkness eons ago. It reveals the magnitude of the energy that slumbers in the nuclei of atoms, and forecasts how many grams of uranium must go into a bomb in order to destroy a city. Finally it discloses some fundamental truths about physical reality. Prior to Relativity scientists had pictured the universe as a vessel containing two distinct elements, matter and energy—the former inert, tangible, and characterized by a property called mass, and the latter active, invisible, and without mass. But Einstein showed that mass and energy are equivalent: the property called mass is simply concentrated energy. In other words matter is energy and energy is matter, and the distinction is simply one of temporary state.

In the light of this broad principle many puzzles of nature are resolved. The baffling interplay of matter and radiation which appears sometimes to be a concourse of particles and sometimes a meeting of waves, becomes more understandable. The dual role of the electron as a unit of matter and a unit of electricity, the wave electron, the photon, waves of matter, waves of probability, a universe of waves—all these seem less paradoxical. For all these concepts simply describe different manifestations of the same underlying reality, and it no longer makes sense to ask what any one of them "really" is. Matter and energy are interchangeable. If matter sheds its mass and travels with the speed of light we call it radiation or energy. And conversely if energy congeals and becomes inert and we can ascertain its mass we call it matter. Heretofore science could only note their ephemeral properties and relations as they touched the perceptions of earthbound man. But since July 16, 1945 man has been able to transform one into the other. For on that night at Alamogordo, New Mexico, man for the first time transmuted a substantial quantity of matter into the light, heat, sound, and motion which we call energy.

*   *   *

Yet the fundamental mystery remains. The whole march of science toward the unification of concepts— the reduction of all matter to elements and then to a few types of particles, the reduction of "forces" to the single concept "energy," and then the reduction of matter *and* energy to a single basic quantity—leads still to the unknown. The many questions merge into one, to which there may never be an answer: what is the essence of this mass-energy substance, what is the underlying stratum of physical reality which science seeks to explore?

Thus Relativity, like the Quantum Theory, draws man's intellect still farther away from the Newtonian universe, firmly rooted in space and time and functioning like some great, unerring, and manageable machine. Einstein's laws of motion, his basic principles of the relativity of distance, time, and mass, and his deductions from these principles comprise what is known as the Special Theory of Relativity. In the decade following the publication of this original work, he expanded this scientific and philosophical system into the General Theory of Relativity, through which he examined the mysterious force that guides the whirling of the stars, comets, meteors, and galaxies, and all the moving systems of iron, stone, vapor, and flame in the immense inscrutable void. Newton called this force "universal gravitation." From his own concept of gravitation Einstein attained a view of the vast architecture and anatomy of the universe as a whole.

9

"THE NONMATHEMATICIAN," SAYS ALBERT Einstein, "is seized by a mysterious shuddering when he hears of 'four-dimensional' things, by a feeling not unlike that awakened by thoughts of the occult. And yet there is no more commonplace statement than that the world in which we live is a four-dimensional space-time continuum."

A nonmathematician might question Einstein's use of the term "commonplace" in this connection. Yet the difficulty lies more in the wording than in ideas. Once the meaning of the word "continuum" is properly grasped Einstein's picture of the universe as a four-dimensional space-time continuum—and this is the view that underlies all modern conceptions of the universe—becomes perfectly clear. A continuum is something that is continuous. A ruler, for example, is a one-dimensional space continuum. Most rulers are divided into inches and fractions, scaled down to one-sixteenth of an inch.

But it is possible to imagine a ruler calibrated to a millionth or a billionth of an inch. In theory there is no reason why the steps from point to point should not be even smaller. The distinguishing characteristic of a continuum is that the interval separating any two points may be divided into an infinite number of arbitrarily small steps.

A railroad track is a one-dimensional space continuum, and on it the engineer of a train can describe his position at any time by citing a single co-ordinate point—i.e., a station or a milestone. A sea captain, however, has to worry about two dimensions. The surface of the sea is a two-dimensional continuum and the co-ordinate points by which a sailor fixes his position in his two-dimensional continuum are latitude and longitude. An airplane pilot guides his plane through a three-dimensional continuum, hence he has to consider not only longitude and latitude, but also his height above the ground. The continuum of an airplane pilot constitutes space as we perceive it. In other words, the space of our world is a three-dimensional continuum.

To describe any physical event involving motion, however, it is not enough simply to indicate position in space. It is necessary to state also how position changes in time. Thus to give an accurate picture of the operation of a New York-Chicago express, one must mention not only that it goes from New York to Albany to Syracuse to Cleveland to Toledo to Chicago, but also the times at which it touches each of those points. This can be done either by means of a timetable or a visual chart. If the miles between New York and Chicago are plotted horizontally on a piece of ruled paper and the hours and minutes are plotted vertically, then a diagonal line properly drawn across the page illustrates the progress of the train in a two-dimensional space-time continuum. This type of graphic representation is familiar to most newspaper readers; a stock market chart, for example, pictures financial events in a two-dimensional dollar-time continuum. In the same way the flight of an airplane from

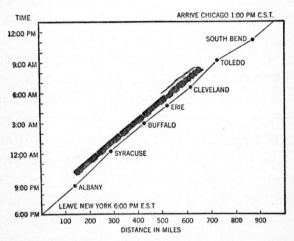

The westbound run of a New York-Chicago express pictured
in a two-dimensional space-time continuum

New York to Los Angeles can best be pictured in a four-dimensional space-time continuum. The fact that the plane is at latitude x, longitude y, and altitude z means nothing to the traffic manager of the airline unless the time co-ordinate is also given. So time is the fourth dimension. And if one wishes to envisage the flight as a whole, as a physical reality, it cannot be broken down into a series of disconnected take-offs, climbs, glides, and landings. Instead it must be thought

63

of as a continuous curve in a four-dimensional space-time continuum.

* * *

Since time is an impalpable quantity it is not possible to draw a picture or construct a model of a four-dimensional space-time continuum. But it can be imagined and it can be represented mathematically. And in order to describe the stupendous reaches of the universe beyond our solar system, beyond the clusters and star clouds of the Milky Way, beyond the lonely outer galaxies burning in the void, the scientist must visualize it all as a continuum in three dimensions of space and one of time. In our minds we tend to separate these dimensions; we have an awareness of space and an awareness of time. But the separation is purely subjective; and as the Special Theory of Relativity showed, space and time separately are relative quantities which vary with individual observers. In any objective description of the universe, such as science demands, the time dimension can no more be detached from the space dimension than length can be detached from breadth and thickness in an accurate representation of a house, a tree, or Betty Grable. According to the great German mathematician, Herman Minkowski, who developed the mathematics of the space-time continuum as a convenient medium for expressing the principles of Relativity, "space and time separately have vanished into the merest shadows, and only a sort of combination of the two preserves any reality."

The Universe and Dr. Einstein

It must not be thought, however, that the space-time continuum is simply a mathematical construction. The world *is* a space-time continuum; all reality exists both in space and in time, and the two are indivisible. All measurements of time are really measurements in space, and conversely measurements in space depend on measurements of time. Seconds, minutes, hours, days, weeks, months, seasons, years, are measurements of the earth's position in space relative to the sun, moon, and stars. Similarly latitude and longitude, the terms whereby man defines his spatial position on earth, are measured in minutes and seconds, and to compute them accurately one must know the time of day and the day of the year. Such "landmarks" as the Equator, the Tropic of Cancer, or the Arctic Circle are simply sundials which clock the changing seasons; the Prime Meridian is a co-ordinate of daily time; and "noon" is nothing more than an angle of the sun.

Even so, the equivalence of space and time becomes really clear only when one contemplates the stars. Among the familiar constellations, some are "real" in that their component stars comprise true gravitational systems, moving in an orderly fashion relative to one another; others are only apparent—their patterns are accidents of perspective, created by a seeming adjacency of unrelated stars along the line of sight. Within such optical constellations one may observe two stars of equal brightness and assert that they are "side by side" in the firmament, whereas in actuality one may be 40 light years and the other 400 light years away.

Obviously the astronomer has to think of the universe as a space-time continuum. When he peers through his telescope he looks not only outward in space but backward in time. His sensitive cameras can detect the glimmer of island universes 500 million light years away—faint gleams that began their journey at a period of terrestrial time when the first vertebrates were starting to crawl from warm Paleozoic seas onto the young continents of Earth. His spectroscope tells him, moreover, that these huge outer systems are hurtling into limbo, away from our own galaxy, at incredible velocities ranging up to 35,000 miles a second. Or, more precisely, they *were* receding from us 500 million years ago. Where they are "now," or whether they even exist "now," no one can say. If we break down our picture of the universe into three subjective dimensions of space and one of local time, then these galaxies have no objective existence save as faint smudges of ancient enfeebled light on a photographic plate. They attain physical reality only in their proper frame of reference, which is the four-dimensional space-time continuum.

*   *   *

In man's brief tenancy on earth he egocentrically orders events in his mind according to his own feelings of past, present, and future. But except on the reels of one's own consciousness, the universe, the objective world of reality, does not "happen"—it simply exists. It can be encompassed in its entire majesty only by a cosmic intellect. But it can also be represented symbolically, by a mathematician, as a four-dimensional space-time continuum. An understanding of the space-time continuum is requisite to a comprehension of the General Theory of Relativity and of what it says about gravitation, the unseen force that holds the universe together and determines its shape and size.

IN THE SPECIAL THEORY OF RELATIVITY, EINSTEIN
studied the phenomenon of motion and showed
that there appears to be no fixed standard in the
universe by which man can judge the "absolute" motion of the earth or of any other moving system. Motion can be detected only as a change of position with
respect to another body. We know for example that
the earth is moving around the sun at the rate of twenty
miles a second. The changing seasons suggest this fact.
But until four hundred years ago men thought the
shifting position of the sun in the sky revealed the
sun's movement around the earth; and on this assumption ancient astronomers developed a perfectly practical system of celestial mechanics which enabled them
to predict with great accuracy all the major phenomena of the heavens. Their supposition was a natural one, for we can't *feel* our motion through space;
nor has any physical experiment ever proved that the
earth actually is in motion. And though all the other
planets, stars, galaxies, and moving systems in the universe are ceaselessly, restlessly changing position, their
movements are observable only with respect to one
another. If all the objects in the universe were re-

moved save one, then no one could say whether that one remaining object was at rest or hurtling through the void at 100,000 miles a second. Motion is a relative state; unless there is some system of reference to which it may be compared, it is meaningless to speak of the motion of a single body.

Shortly after publishing the Special Theory of Relativity, however, Einstein began wondering if there is not indeed one kind of motion which may be considered "absolute" in that it can be detected by the physical effect it exerts on the moving system itself— without reference to any other system. For example, an observer in a *smoothly* running train is unable to tell by experiments performed inside the train whether he is in motion or at rest. But if the engineer of the train suddenly applies the brakes or jerks open the throttle, he will then be made aware, by the resulting jolt, of a change in his velocity. And if the train rounds a turn, he will know by the outward tug of his own body, resisting a change of direction, that the train's course has been altered in a certain way. Therefore, Einstein reasoned, if only one object existed in the entire universe—the earth for example—and it suddenly began to gyrate irregularly, its inhabitants would be uncomfortably aware of their motion. This suggests that nonuniform motion, such as that produced by forces and accelerations, may be "absolute" after all. It also suggests that empty space can serve as a system of reference within which it is possible to distinguish absolute motion.

To Einstein, who held that space is nothing and motion is relative, the apparently unique character of nonuniform motion was profoundly disturbing. In the Special Theory of Relativity he had taken as his premise the simple assertion that the laws of nature are the same for all systems moving *uniformly* relative to one another. And as a steadfast believer in the universal harmony of nature he refused to believe that any system in a state of nonuniform motion must be a uniquely distinguished system in which the laws of nature are different. Hence as the basic premise of his General Theory of Relativity, he stated: the laws of nature are the same for all systems regardless of their state of motion. In developing this thesis he worked out new laws of gravitation which upset most of the concepts that had shaped man's picture of the universe for three hundred years.

*　　　*　　　*

Einstein's springboard was Newton's Law of Inertia which, as every schoolboy knows, states that "every body continues in its state of rest, or of uniform motion in a straight line, unless it is compelled to change that state by forces impressed thereon." It is inertia, therefore, which produces our peculiar sensations when a railroad train suddenly slows down or speeds up or rounds a curve. Our body wants to continue moving uniformly in a straight line, and when the train impresses an opposing force upon us the property called inertia tends to resist that force. It is also inertia which causes a locomotive to wheeze and strain in order to accelerate a long train of freight cars.

But this leads to another consideration. If the cars are loaded the locomotive has to work harder and burn more coal than if the cars are empty. To his Law of Inertia Newton therefore added a second law stating that the amount of force necessary to accelerate a body depends on the mass of the body; and that if the same force is applied to two bodies of different masses, then it will produce a greater acceleration in the smaller body than in the bigger one. This principle holds true for the whole range of man's everyday experience— from pushing a baby carriage to firing a cannon. It simply generalizes the obvious fact that one can throw a baseball farther and faster than one can throw a cannonball.

There is, however, one peculiar situation in which there appears to be no connection between the acceleration of a moving body and its mass. The baseball and the cannonball attain exactly the same rate of acceleration when they are *falling*. This phenomenon was first discovered by Galileo, who proved by experiment that, discounting air resistance, bodies all fall at precisely the same rate regardless of their size or composition. A baseball and a handkerchief fall at different speeds only because the handkerchief offers a larger surface to the resistance of the air. But objects of comparable shape, such as a marble, a baseball, and a cannonball, fall at the same rate. (In a vacuum the handkerchief and the cannonball would fall side by side.)

This phenomenon appears to violate Newton's Law of Inertia. For why should all objects travel vertically at the same velocity regardless of their size or mass, if those same objects, when projected horizontally by an equal force, move at velocities that are strictly determined by their mass? It would appear as though the factor of inertia operates only in a horizontal plane.

Newton's solution of this riddle is given in his Law of Gravitation, which states simply that the mysterious force by which a material body attracts another body increases with the mass of the object it attracts. The bigger the object, the stronger the call of gravity. If an object is small, its inertia or tendency to resist motion is small, but the force that gravity exerts upon it is also small. If an object is big, its inertia is great, but the force that gravity exerts upon it is also great. Hence gravity is always exerted in the precise degree necessary to overcome the inertia of any object. And that is why all objects fall at the same rate, regardless of their inertial mass.

This rather remarkable coincidence—the perfect balance of gravitation and inertia—was accepted on faith, but never understood or explained, for three centuries after Newton. All of modern mechanics and engineering grew out of Newtonian concepts, and the heavens appeared to operate in accordance with his laws. Einstein, however, whose discoveries have all sprung from an inherent distrust of dogma, disliked several of Newton's assumptions. He doubted that the balance of gravitation and inertia was merely an accident of nature. And he rejected the idea of gravitation being a force that can be exerted instantaneously over great distances. The notion that the earth can reach out into space and pull an object toward it with a force miraculously and invariably equal to the inertial resistance of that object seemed to Einstein highly improbable. So out of his objections he evolved a new theory of gravitation which, experience has shown, gives a more accurate picture of nature than Newton's classic law.

IN ACCORDANCE WITH HIS USUAL MODE OF CREA-
tive thought Einstein set the stage with an imaginary situation. The details have doubtless been
envisaged by many another dreamer in restless slum-
ber or in moments of insomniac fancy. He pictured an
immensely high building and inside it an elevator that
had slipped from its cables and is falling freely. Within the elevator a group of physicists, undisturbed by
any suspicion that their ride might end in disaster, are
performing experiments. They take objects from their
pockets, a fountain pen, a coin, a bunch of keys, and
release them from their grasp. Nothing happens. The
pen, the coin, the keys appear to the men in the elevator
to remain poised in mid-air—because all of them are
falling, along with the elevator and the men, at precisely the same rate in accordance with Newton's Law
of Gravitation. Since the men in the elevator are
unaware of their predicament, however, they may
explain these peculiar happenings by a different assumption. They may believe they have been magically
transported outside the gravitational field of the earth
and are in fact poised somewhere in empty space. And
they have good grounds for such a belief. If one of them

jumps from the floor he floats smoothly toward the ceiling with a velocity just proportional to the vigor of his jump. If he pushes his pen or his keys in any direction, they continue to move uniformly in that direction until they hit the wall of the car. Everything apparently obeys Newton's Law of Inertia, and continues in its state of rest or of uniform motion in a straight line. The elevator has somehow become an inertial system, and there is no way for the men inside it to tell whether they are falling in a gravitational field or are simply floating in empty space, free from all external forces.

Einstein now shifts the scene. The physicists are still in the elevator, but this time they really *are* in empty space, far away from the attractive power of any celestial body. A cable is attached to the roof of the elevator; some supernatural force begins reeling in the cable; and the elevator travels "upward" with constant acceleration—i.e., progressively faster and faster. Again the men in the car have no idea where they are, and again they perform experiments to evaluate their situation. This time they notice that their feet press solidly against the floor. If they jump they do not float to the ceiling for the floor comes up beneath them. If they release objects from their hands the objects appear to "fall." If they toss objects in a horizontal direction they do not move uniformly in a straight line but describe a parabolic curve with respect to the floor. And so the scientists, who have no idea that their windowless car actually is climbing through interstellar space, conclude that they are situated in quite ordinary circumstances in a stationary room rigidly attached to the earth and affected in normal measure by the force of gravity. There is really no way for them to tell whether they are at rest in a gravitational field or ascending with constant acceleration through outer space where there is no gravity at all.

The same dilemma would confront them if their room were attached to the rim of a huge rotating merry-go-round set in outer space. They would feel a strange force trying to pull them away from the center of the merry-go-round, and a sophisticated outside observer would quickly identify this force as inertia (or, as it is termed in the case of rotating objects, centrifugal force). But the men inside the room, who as usual are unaware of their odd predicament, would once again attribute the force to gravity. For if the interior of their room is empty and unadorned, there will be nothing to tell them which is the floor and which is the ceiling except the force that pulls them toward one of its interior surfaces. So what a detached observer would call the "outside wall" of the rotating room becomes the "floor" of the room for the men inside. A moment's reflection shows that there is no "up" or "down" in empty space. What we on earth call "down" is simply the direction of gravity. To a man on the sun it would appear that the Australians, Africans, and Argentines are hanging by their heels from the southern hemisphere. By the same token Admiral Byrd's flight over the South Pole was a geometrical fiction; actually he flew *under* it—upside down. And so the men inside the room on the merry-go-round will find that all their experiments produce exactly the same results as the ones they performed when their room was being swept "upward" through space. Their feet stay firmly on the "floor." Solid objects "fall." And once again they attribute these phenomena to the force of gravity and believe themselves at rest in a gravitational field.

*   *   *

From these fanciful occurrences Einstein drew a conclusion of great theoretical importance. To physicists it is known as the Principle of Equivalence of Gravitation and Inertia. It simply states that there is no way to distinguish the motion produced by inertial forces (acceleration, recoil, centrifugal force, etc.) from motion produced by gravitational force. The validity of this principle will be evident to any aviator; for in an airplane it is impossible to separate the effects of inertia from those of gravitation. The physical sensation of pulling out of a dive is exactly the same as that produced by executing a steeply banked turn at high speed. In both cases the factor known to flyers as a "G-load" (Gravity load) appears, blood is drawn away from the head, and the body is pulled heavily down into the seat. But in one case the effects are produced by gravity, in the other by inertia.

In this principle, which is the keystone of General Relativity, Einstein found an answer both to the riddle of gravitation and the problem of "absolute" motion.

The Universe and Dr. Einstein

It showed that there is nothing unique or "absolute" about nonuniform motion after all; for the effects of nonuniform motion which can supposedly reveal the state of motion of a body, even if it exists alone in space, are indistinguishable from the effects of gravitation. Thus in the case of the merry-go-round, what one observer identified as the pull of inertia or centrifugal force and therefore an effect of motion, another observer identified as the familiar tug of gravitation. And any other inertial effect produced by a change of speed or a change of direction can equally well be ascribed to a changing or fluctuating gravitational field. So the basic premise of Relativity holds true; motion, both uniform and nonuniform, can only be judged with respect to some system of reference—absolute motion does not exist.

The sword with which Einstein slew the dragon of absolute motion was gravitation. But what *is* gravitation? The gravitation of Einstein is something entirely different from the gravitation of Newton. It is not a "force." The idea that bodies of matter can "attract" one another is, according to Einstein, an illusion that has grown out of erroneous mechanical concepts of nature. So long as one believes that the universe is a big machine, it is natural to think that its various parts can exert a force on one another. But the deeper science probes toward reality, the more clearly it appears that the universe is not like a machine at all. So Einstein's Law of Gravitation contains nothing about force. It describes the behavior of objects in a gravitational field

—the planets, for example—not in terms of "attraction" but simply in terms of the paths they follow. To Einstein, gravitation is simply part of inertia; the movements of the stars and the planets arise from their inherent inertia; and the courses they follow are determined by the metric properties of space—or more properly speaking, the metric properties of the space-time continuum.

Although this sounds very abstract and even paradoxical, it becomes quite clear as soon as one dismisses the notion that bodies of matter can exert a physical force on each other across millions of miles of empty space. This concept of "action-at-a-distance" has troubled scientists since Newton's day. It led to particular difficulty, for example, in understanding electric and magnetic phenomena. Today scientists no longer say that a magnet "attracts" a piece of iron by some kind of mysterious but instantaneous action-at-a-distance. They say rather that the magnet creates a certain physical condition in the space around it, which they term a magnetic field; and that this magnetic field then acts upon the iron and makes it behave in a certain predictable fashion. Students in any elementary science course know what a magnetic field looks like, because it can be rendered visible by the simple process of shaking iron filings onto a piece of stiff paper held above a magnet. A magnetic field and an electrical field are physical realities. They have a definite structure, and their structure is described by the field equations of James Clerk Maxwell which pointed the way toward all the discoveries in electrical and radio engineering of the past century. A gravitational field is as much of a physical reality as an electromagnetic field, and its structure is defined by the field equations of Albert Einstein.

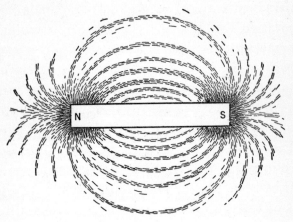

The field of a bar magnet

* * *

Just as Maxwell and Faraday assumed that a magnet creates certain properties in surrounding space, so Einstein concluded that stars, moons, and other celestial objects individually determine the properties of the space around them. And just as the movement of a piece of iron in a magnetic field is guided by the structure of the field, so the path of any body in a gravitational field is determined by the geometry of that field. The distinction between Newton's and Einstein's ideas about gravitation has sometimes been illustrated by picturing a little boy playing marbles in a city lot. The ground is very uneven, ridged with bumps and hollows. An observer in an office ten stories above the street would not be able to see these irregularities in the ground. Noticing that the marbles appear to avoid some sections of the ground and move toward other sections, he might assume that a "force" is operating which repels the marbles from certain spots and attracts them toward others. But another observer on the ground would instantly perceive that the path of the marbles is simply governed by the curvature of the field. In this little fable Newton is the upstairs observer who imagines that a "force" is at work, and Einstein is the observer on the ground, who has no reason to make such an assumption. Einstein's gravitational laws, therefore, merely describe the field properties of the space-time continuum. Specifically, one group of these laws sets forth the relation between the mass of a gravitating body and the structure of the field around it; they are called structure laws. A second group analyzes the paths described by moving bodies in gravitational fields; they are the laws of motion.

It should not be thought that Einstein's theory of gravitation is only a formal mathematical scheme. For it rests on assumptions of deep cosmic significance. And the most remarkable of these assumptions is that the universe is not a rigid and immutable edifice where independent matter is housed in independent space and time; it is on the contrary an amorphous continuum, without any fixed architecture, plastic and variable, constantly subject to change and distortion. Wherever there is matter and motion, the continuum is disturbed. Just as a fish swimming in the sea agitates the water around it, so a star, a comet, or a galaxy distorts the geometry of the space-time through which it moves.

When applied to astronomical problems Einstein's gravitational laws yield results that are close to those given by Newton. If the results paralleled each other in every case, scientists might tend to retain the familiar concepts of Newtonian law and write off Einstein's theory as a weird if original fancy. But a number of strange new phenomena have been discovered, and at least one old puzzle solved, solely on the basis of General Relativity. The old puzzle stemmed from the eccentric behavior of the planet Mercury. Instead of revolving in its elliptical orbit with the regularity of the other planets, Mercury deviates from its course each year by a slight but exasperating degree. Astronomers explored every possible factor that might cause this perturbation but found no solution within the framework of Newtonian theory. It was not until Einstein evolved his laws of gravitation that the problem was solved. Of all the planets Mercury lies closest to the sun. It is small and travels with great speed. Under Newtonian law these factors should not in themselves account for the deviation; the dynamics of Mercury's movement should be basically the same as those of any other planet. But under Einstein's laws, the intensity of the sun's gravitational field and Mercury's enormous speed make a difference, causing the whole ellipse of Mercury's orbit to execute a slow but inexorable swing around the sun at the rate of one revolution in 3,000,000 years. This calculation is in perfect agreement with actual measurements of the planet's course. Einstein's mathematics are thus more accurate than Newton's in dealing with high velocities and strong gravitational fields.

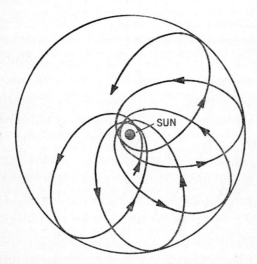

The rotation of Mercury's elliptical orbit, greatly exaggerated. Actually the ellipse advances only 43 seconds of an arc per century.

An achievement of far greater importance, however, than this solution of an old problem was Einstein's prediction of a new cosmic phenomenon of which no scientist had ever dreamed—namely the effect of gravitation on light.

THE SEQUENCE OF THOUGHT WHICH LED EIN-stein to prophesy this effect began with an-other imaginary situation. As before, the scene opens in an elevator ascending with constant accelera-tion through empty space, far from any gravitational field. This time some roving interstellar gunman im-pulsively fires a bullet at the elevator. The bullet hits the side of the car, passes clean through and emerges from the far wall at a point a little below the point at which it penetrated the first wall. The reason for this is evident to the marksman on the outside. He knows that the bullet flew in a straight line, obeying Newton's Law of Inertia; but while it traversed the distance be-tween the two walls of the car, the whole elevator trav-eled "upward" a certain distance, causing the second bullet hole to appear not opposite the first one but slightly nearer the floor. However the observers inside the elevator, having no idea where in the universe they are, interpret the situation differently. Aware that on earth any missile describes a parabolic curve toward the ground, they simply conclude that they are at rest in a gravitational field and that the bullet which passed through their car was describing a perfectly normal curve with respect to the floor.

A moment later as the car continues upward through space a beam of light is suddenly flashed through an aperture in the side of the car. Since the velocity of light is great the beam traverses the distance between its point of entrance and the opposite wall in a very small fraction of a second. Nevertheless the car travels upward a certain distance in that interval, so the beam strikes the far wall a tiny fraction of an inch below the point at which it entered. If the observers within the car are equipped with sufficiently delicate instruments of measurement they will be able to compute the curvature of the beam. But the question is, how will they explain it? They are still unaware of the motion of their car and believe themselves at rest in a gravitational field. If they cling to Newtonian principles they will be completely baffled because they will insist that light rays always travel in a straight line. But if they are familiar with the Special Theory of Relativity they will remember that energy has mass in accordance with the equation $m = E/c^2$. Since light is a form of energy they will deduce that light has mass and will therefore be affected by a gravitational field. Hence the curvature of the beam.

From these purely theoretical considerations Einstein concluded that light, like any material object, travels in a curve when passing through the gravitational field of a massive body. He suggested that his theory could be put to test by observing the path of starlight in the gravitational field of the sun. Since the stars are invisible by day, there is only one occasion when sun and stars can be seen together in the sky, and that is during an eclipse. Einstein proposed, therefore, that photographs be taken of the stars immediately

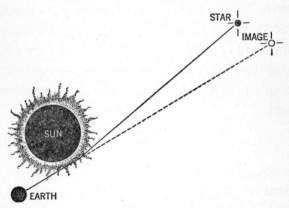

The deflection of starlight in the gravitational field of the sun. Since the light from a star in the neighborhood of the sun's disk is bent inward, toward the sun, as it passes through the sun's gravitational field, the image of the star appears to observers on earth to be shifted outward and away from the sun.

bordering the darkened face of the sun during an eclipse and compared with photographs of those same stars made at another time. According to his theory the light from the stars surrounding the sun should be bent inward, toward the sun, in traversing the sun's gravitational field; hence the *images* of those stars should appear to observers on earth to be shifted outward from their usual positions in the sky. Einstein calculated the degree of deflection that should be observed and predicted that for the stars closest to the sun the deviation would be about 1.75 seconds of an arc. Since he staked his whole General Theory of Relativity on this test, men of science throughout the world anxiously awaited the findings of expeditions which journeyed to equatorial regions to photograph the eclipse of May 29, 1919. When their pictures were developed and examined, the deflection of the starlight in the gravitational field of the sun was found to average 1.64 seconds—a figure as close to perfect agreement with Einstein's prediction as the accuracy of instruments allowed.

* * *

Another prediction made by Einstein on the basis of General Relativity pertained to time. Having shown how the properties of space are affected by a gravitational field, Einstein reached the conclusion by analogous but somewhat more involved reasoning that time intervals also vary with the gravitational field. A clock transported to the sun should run at a slightly slower rhythm than on earth. And a radiating solar atom should emit light of slightly lower frequency than an atom of the same element on earth. The difference in wave length would in this case be immeasurably small. But there are in the universe gravitational fields stronger than the sun's. One of these surrounds the freak star known as the "companion of Sirius"—a white dwarf composed of matter in a state of such fantastic density that one cubic inch of it would weigh a ton on earth. Because of its great mass this extraordinary dwarf, which is only three times larger than the earth, has a gravitational field potent enough to perturb the movements of Sirius, seventy times its size. Its field is also powerful enough to slow down the frequency of its own radiation by a measurable degree, and spectroscopic observations have indeed proved that the frequency of light emitted by Sirius' companion is reduced by the exact amount predicted by Einstein. The shift of wave length in the spectrum of this star is known to astronomers as "the Einstein Effect" and constitutes an additional verification of General Relativity.

U P TO THIS POINT THE CONCEPTS OF GENERAL
Relativity have dealt with the phenomena of
the individual gravitational field. But the universe is filled with incomputable bodies of matter—
meteors, moons, comets, nebulae, and billions on billions of stars grouped by the interlocking geometry
of their gravitational fields in clusters, clouds, galaxies,
and supergalactic systems. One naturally asks, what
then is the over-all geometry of the space-time continuum in which they drift? In cruder language, what
is the shape and size of the universe? All modern re-
plies to the question have been derived directly or in-
directly from the principles of General Relativity.

Prior to Einstein the universe was most commonly
pictured as an island of matter afloat in the center of
an infinite sea of space. There were several reasons for
this concept. The universe, most scientists agreed, had
to be infinite; because as soon as they conceded that
space might come to an end somewhere, they were
faced with the embarrassing question: "And what lies
beyond that?" Yet Newtonian law prohibited an infinite universe containing a uniform distribution of
matter, for then the total gravitational force of all the

masses of matter stretching away to infinity would be infinite, and the heavens would be ablaze with infinite light. To man's feeble eye, moreover, it appeared that beyond the rim of our Milky Way the lamps of space became sparser and sparser, diffusing gradually in attenuated outposts like lonely lighthouses on the frontiers of the fathomless void. But the island universe presented difficulties too. The amount of matter it held was so small by contrast with an infinity of space that inevitably the dynamic laws governing the movements of the galaxies would cause them to disperse like the droplets of a cloud and the universe would become entirely empty.

To Einstein this picture of dissolution and disappearance seemed eminently unsatisfactory. The basic difficulty, he decided, derived from man's natural but unwarranted assumption that the geometry of the universe must be the same as that revealed by his senses here on earth. We confidently assume, for example, that two parallel beams of light will travel through space forever without meeting, because in the infinite plane of Euclidean geometry parallel lines never meet. We also feel certain that in outer space, as on a tennis court, a straight line is the shortest distance between two points. And yet Euclid never actually *proved* that a straight line is the shortest distance between two points; he simply arbitrarily *defined* a straight line as the shortest distance between two points.

Is it not then possible, Einstein asked, that man is being deceived by his limited perceptions when he pictures the universe in the garb of Euclidean geometry? There was a time when man thought the earth was flat. Now he accepts the fact that the earth is round, and he knows that on the surface of the earth the shortest distance between two such points as New York and London is not a straight line across the Atlantic but a "great circle" that veers northward past Nova Scotia, Newfoundland, and Iceland. So far as the surface of the earth is concerned Euclid's geometry is not valid. A giant triangle, drawn on the earth's surface from two points on the Equator to the North Pole would not satisfy Euclid's theorem that the sum of the interior angles of a triangle is always equal to two right angles or 180 degrees. It would contain *more* than 180 degrees, as a glance at the globe will quickly show. And if someone should draw a giant circle on the earth's surface he would find that the ratio between its diam-

eter and its circumference is less than the classic value
pi. These departures from Euclid are due to the curvature of the earth. Although no one doubts today that
the earth has a curvature, man did not discover this
fact by getting off the earth and looking at it. The curvature of the earth can be computed very comfortably
on terra firma by a proper mathematical interpretation
of easily observable facts. In the same way, by a synthesis of astronomical fact and deduction, Einstein
concluded that the universe is neither infinite nor
Euclidean, as most scientists supposed, but something
hitherto unimagined.

*   *   *

It has already been shown that Euclidean geometry
does not hold true in a gravitational field. Light rays
do not travel in straight lines when passing through a
gravitational field, for the geometry of the field is such
that within it there are no straight lines; the shortest
course that the light can describe is a curve or great
circle which is rigorously determined by the geometrical structure of the field. Since the structure of a gravitational field is shaped by the mass and velocity of
the gravitating body—star, moon, or planet—it follows that the geometrical structure of the universe as
a whole must be shaped by the sum of its material content. For each concentration of matter in the universe
there is a corresponding distortion of the space-time
continuum. Each celestial body, each galaxy creates
local irregularities in space-time, like eddies around

islands in the sea. The greater the concentration of matter, the greater the resulting curvature of space-time. And the total effect is an over-all curvature of the whole space-time continuum: the combined distortions produced by all the incomputable masses of matter in the universe cause the continuum to bend back on itself in a great closed cosmic curve.

The Einstein universe therefore is non-Euclidean and finite. To earthbound man a light ray may appear to travel in a straight line to infinity, just as to an earthworm crawling "straight" ahead forever and ever the earth may seem both flat and infinite. But man's impression that the universe is Euclidean in character, like the earthworm's impression of the earth, is imparted by the limitations of his senses. In the Einstein universe there are no straight lines, there are only great circles. Space, though finite, is unbounded; a mathematician would describe its geometrical character as the four-dimensional analogue of the surface of a sphere. In the less abstract words of the late British physicist, Sir James Jeans:

"A soap-bubble with corrugations on its surface is perhaps the best representation, in terms of simple and familiar materials, of the new universe revealed to us by the Theory of Relativity. The universe is not the interior of the soap-bubble but its surface, and we must always remember that while the surface of the soap-bubble has only two dimensions, the universe bubble has four—three dimensions of space and one of time.

And the substance out of which this bubble is blown, the soap-film, is empty space welded onto empty time."

Like most of the concepts of modern science, Einstein's finite, spherical universe cannot be visualized— any more than a photon or an electron can be visualized. But as in the case of the photon and the electron its properties can be described mathematically. By taking the best available values of modern astronomy and applying them to Einstein's field equations, it is possible to compute the *size* of the universe. In order to determine its radius, however, it is first necessary to ascertain its curvature. Since, as Einstein showed, the geometry or curvature of space is determined by its material content, the cosmological problem can be solved only by obtaining a figure for the average density of matter in the universe.

Fortunately this figure is available, for astronomer Edwin Hubble of the Mt. Wilson Observatory conscientiously studied sample areas of the heavens over a period of years and painstakingly computed the average amount of matter contained in them. The conclusion he reached was that in the universe as a whole there is .0000000000000000000000000000001 gram of matter per cubic centimeter of space. Applied to Einstein's field equations this figure yields a positive value for the curvature of the universe, which in turn reveals that the radius of the universe is 35 billion light years or 210,000,000,000,000,000,000,000,000 miles. Einstein's universe, while not infinite, is nevertheless sufficiently enormous to encompass billions of galaxies, each containing hundreds of millions of flaming stars and incalculable quantities of rarefied gas, cold systems of iron and stone and cosmic dust. A sunbeam, setting out through space at the rate of 186,000 miles a second would, in this universe, describe a great cosmic circle and return to its source after a little more than 200 billion terrestrial years.

14

A T THE TIME EINSTEIN EVOLVED HIS COSMOLOGY, he was unaware, however, of a strange astronomical phenomenon which was only interpreted several years later. He had assumed that the motions of the stars and galaxies were random, like the aimless drifting of molecules in a gas. Since there was no evidence of any unity in their wanderings he had ignored them entirely and regarded the universe as static. But astronomers were beginning to notice signs of a systematic movement among the outer galaxies at the extreme limits of telescopic vision. All these outlying galaxies, or "island universes," are, apparently, receding from our solar system and from each other. This organized flight of the distant galaxies— the remotest of them being about 500 million light years away—is an entirely different affair from the indolent wheeling of the nearer gravitational systems. For such a systematic movement would have an effect on the curvature of the universe as a whole.

The universe is, therefore, not static; it is expanding in somewhat the same manner as a soap bubble or a balloon expands. The analogy is not quite exact, however, for if we conceive of the universe as a kind of

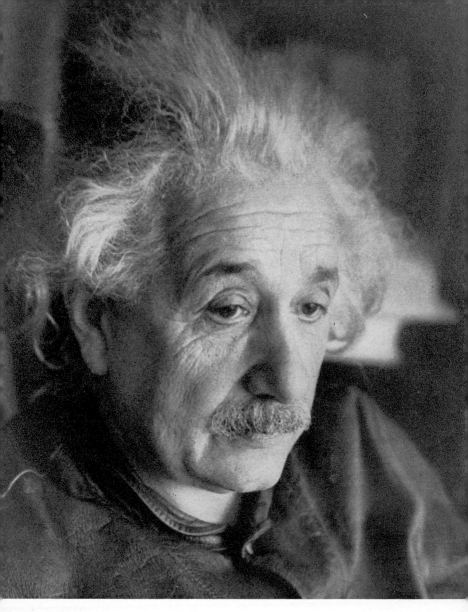

DR. ALBERT EINSTEIN

Yerkes Observatory 10-inch, Bruce Refractor

The Milky Way which girdles our heavens with a band of pearly light is actually a vast ocean of suns, star fields, clusters, and clouds, composing the galaxy within which our immediate star system moves. Our sun, far from being the center of the universe, is not even the center of our galaxy, but rides on the periphery 30,000 light years from the galactic nucleus. The dark areas suggesting rents in the stellar canopy through which one peers into the void, are actually obscuring masses of opaque matter, dust clouds, whose significance is discussed on pages 101-2.

Lick Observatory, Crossley Reflector

The great Andromeda galaxy is a giant star system similar in shape and structure to our own Milky Way. Although it can be seen with the naked eye as a faint luminescence in the constellation Andromeda, it is 700,000 light years away. Yet it is the nearest of all the island universes that wheel in the depths of space. Its diameter is 60,000 light years. To an observer situated in this galaxy, our Milky Way would look very much like this. The smaller nebulosities near by are minor members of the super-galactic cluster that encompasses the Andromeda spiral, our Milky Way, and the Magellanic Clouds.

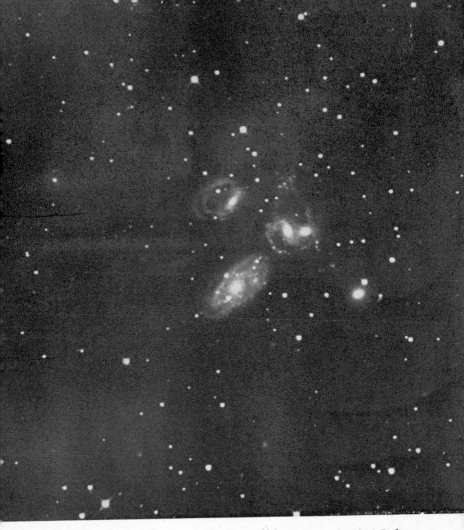

Mount Wilson Observatory, 60-inch Reflector

The Pegasus quintet comprises a galactic supersystem—a cluster of galaxies—each member an island universe composed, like our Milky Way, of hundreds of millions of stars. Note how the interpenetrating effects of gravitation have distorted several of its members. So deeply sunk in space are these galaxies that the light by which this photograph was made took 22 million years to traverse the stupendous reaches between its origins and earth. The immensity of the universe can be comprehended only when one realizes that as our solar system is to the Milky Way, so our Milky Way is to the systems of outer space. For an account of what these remote galaxies tell astronomers about the cosmos, see page 96 ff.

spotted balloon—the spots representing matter—one would expect the spots to expand too. But this cannot be, because then we would never notice the expansion, just as Alice in Wonderland would have been unaware of her sudden changes in stature if all her surroundings had grown and contracted along with her. Therefore, as cosmologist H. P. Robertson of the California Institute of Technology has pointed out, in visualizing the universe as a spotted balloon, we must think of the spots as inelastic patches sewn upon the surface. Material bodies retain their dimensions while space stretches out between them, like the skin of the balloon between the patches.

This extraordinary phenomenon greatly complicates cosmology. If the spectroscopic analysis that indicates the recession of these outer galaxies is correct (as most astronomers believe it to be) then the velocities at which they are vanishing into limbo are almost beyond belief. Their speed appears to increase with distance. While the nearer galaxies, about one million light years away, are traveling at a mere 100 miles a second, those 250 million light years away are flying off at the fantastic rate of 25,000 miles a second, almost one seventh the velocity of light. Since all of these remote galaxies, without exception, are moving *away* from us and from each other, one must conclude that at some epoch of cosmic time all of them were clustered together in one fiery inchoate mass. And if the geometry of space is shaped by its material content then the universe in this pregalactic phase must have

been an uncomfortably cramped and crowded recep-
tacle, characterized by an excessive curvature and
packed with matter in a state of inconceivable density.
Calculations based on the velocities of the receding
galaxies show that they must have separated and
started their flight from the "center" of this shrunken
universe about two billion years ago.

*　　*　　*

Several theories have been advanced by astronomers
and cosmologists to explain the enigma of the expanding universe. One, put forth by the Belgian cosmologist, Abbé Lemaître, proposes that the universe originated from a single stupendous primordial atom which
exploded and thus precipitated the expansion which
we still perceive. An analogous theory, made public
recently by Dr. George Gamow of George Washing-
ton University, reconstructs in detail how the con-
stituent elements might have been forged in the dense
flaming core of the universe before it started to expand. In the beginning, says Dr. Gamow, his nucleus
of the universe was an inferno of homogeneous primordial vapor seething at unimaginable temperatures
such as no longer exist even in the interiors of stars.
(The temperature of the sun, which is an average star,
ranges from $5500°$ Centigrade at the surface up to $40,-
000,000°$ in the interior.) There were no elements in
such heat, no molecules, no atoms—nothing but free
neutrons in a state of chaotic agitation. When the cos-

mic mass began to expand, however, the temperature began to fall; and when it had dropped to about one million degrees the neutrons condensed into aggregates; electrons were emitted which attached themselves to nuclei, and atoms were formed. All the elements in the universe were thus created within the space of a few critical moments in the cosmic dawn and their roles fixed for the two billion years of continuing expansion that ensued.

An earlier theory of the expanding universe, put forth some years ago by Dr. R. C. Tolman of the California Institute of Technology, suggests that the cosmic expansion may be simply a temporary condition which will be followed at some future epoch of cosmic time by a period of contraction. The universe in this picture is a pulsating balloon in which cycles of expansion and contraction succeed each other through eternity. These cycles are governed by changes in the amount of matter in the universe; for as Einstein showed, the curvature of the universe is dependent on its content. The difficulty with this theory is that it rests on the assumption that somewhere in the universe matter is being formed. Although it is true that the amount of matter in the universe is perpetually changing, the change appears to be all in one direction—toward dissolution. All the phenomena of nature, visible and invisible, within the atom and in outer space, indicate that the substance and energy of the universe are inexorably diffusing like vapor through the insatiable void. The sun is slowly but surely burning out, the stars are dying embers, and everywhere in the cosmos heat is turning to cold, matter is dissolving into radiation, and energy is being dissipated into empty space.

The universe is thus progressing toward an ultimate "heat-death," or as it is technically defined, a condition of "maximum entropy." When the universe reaches this state some billions of years from now all the processes of nature will cease. All space will be at the same temperature. No energy can be used because all of it will be uniformly distributed through the cosmos. There will be no light, no life, no warmth—nothing but perpetual and irrevocable stagnation. Time itself will come to an end. For entropy points the direction of time. Entropy is the measure of randomness. When all system and order in the universe have vanished, when randomness is at its maximum, and entropy cannot be increased, when there no longer is any sequence of cause and effect, in short when the universe has run down, there will be no direction to time—there will be no time. And there is no way of avoiding this destiny. For the fateful principle known as the Second Law of Thermodynamics, which stands today as virtually the only pillar of classical physics left intact by the march of science, proclaims that the fundamental processes of nature are irreversible. Nature moves just one way.

There are a few contemporary theorists, however, who propose that somehow, somewhere beyond man's meager ken the universe may be rebuilding itself. In the light of Einstein's principle of the equivalence of mass and energy, it is possible to imagine the diffused radiation in space congealing once more into particles of matter—electrons, atoms, and molecules—which may then combine to form larger units, which in turn may be collected by their own gravitational influence into diffuse nebulae, stars, and, ultimately, galactic systems. And thus the life cycle of the universe may be repeated for all eternity. Laboratory experiments have indeed demonstrated that photons of high-energy radiation, such as gamma rays, can, under certain conditions, interact with matter to produce pairs of electrons and positrons. Astronomers have also determined recently that atoms of the lighter elements, drifting in space—hydrogen, helium, oxygen, nitrogen, and carbon—may slowly coalesce into molecules and microscopic particles of dust and gas. And still more recently Dr. Fred L. Whipple of Harvard has described in his "Dust Cloud Hypothesis", published in 1948, how the rarefied cosmic dust that floats in interstellar space in quantities equal in mass to all the visible matter in the universe could in the course of a billion years condense and coagulate into stars. According to Whipple these tiny dust particles, barely one fifty-thousandths of an inch in diameter, are blown together by the delicate pressure of starlight, just as the fine-spun tail of a comet is deflected away from the sun by the impact of solar photons. As the particles cohere, an aggregate is formed, then a cloudlet, and then a cloud. When the cloud attains gigantic

proportions (i.e., when its diameter exceeds six trillion miles), its mass and density will be sufficient to
set a new sequence of physical processes into operation. Gravity will cause the cloud to contract, and its
contraction will cause its internal pressure and temperature to rise. Eventually, in the last white-hot
stages of its collapse, it will begin to radiate as a star.
Theory shows that our solar system might have
evolved, in special circumstances, from such a process
—our sun being the star in question and the various
planets small cold by-products condensed from sub-
sidiary cloudlets spiraling within the main cloud.

Presupposing the possibility of such events as these,
one might arrive ultimately at the concept of a self-
perpetuating pulsating universe, renewing its cycles
of formation and dissolution, light and darkness,
order and disorder, heat and cold, expansion and contraction, through never-ending eons of time. And yet
this picture has not been widely accepted because no
definitive evidence has been found to support it.
Although dust clouds of all dimensions and degrees
of density can be seen hanging in the abyss of interstellar space, no one can state from man's brief tem-
poral perspective that they are proto-stars, any more
than one can say with assurance that a white cumulus
cloud riding the blue atmosphere of our earth on any
given day is tomorrow's thunder storm or simply an
evanescent wraith of mist that winds have gathered
and will presently disperse. But apart from conjecture on the origins of our solar system or the individual

stars or any component part of the system of nature in which we stand, there are theoretical as well as empirical difficulties inherent in every suggestion that the universe as a whole may still be abuilding. Nothing in all inanimate nature can be unmistakably identified as a pure creative process. At one time, for example, it was thought that the mysterious cosmic rays which continually bombard the earth from outer space might be by-products of some process of atomic creation. But there is greater support for the opposite view that they are by-products of atomic annihilation. Everything indeed, everything visible in nature or established in theory, suggests that the universe is implacably progressing toward final darkness and decay.

There is an important philosophical corollary to this view. For if the universe is running down and nature's processes are proceeding in just one direction, the inescapable inference is that everything had a *beginning*: somehow and sometime the cosmic processes were started, the stellar fires ignited, and the whole vast pageant of the universe brought into being. Most of the clues, moreover, that have been discovered at the inner and outer frontiers of scientific cognition suggest a definite time of Creation. The unvarying rate at which uranium expends its nuclear energies and the absence of any natural process leading to its formation indicate that all the uranium on earth must have come into existence at one specific time, which, according to the best calculations of geophysicists, was about two billion years ago. The tempo at which the wild thermonuclear processes in the interiors of stars transmute matter into radiation enables astronomers to compute with fair assurance the duration of stellar life, and the figure they reach as the likely average age of most stars visible in the firmament today is two billion years. The arithmetic of the geophysicists and astrophysicists is thus in striking agreement with that of the cosmogonists who, basing their calculations on the apparent velocity of the receding galaxies, find that the universe began to expand two billion years ago. And there are other signs in other areas of science that submit the same reckoning. So all the evidence that points to the ultimate annihilation of the universe points just as definitely to an inception fixed in time.

Even if one acquiesces to the idea of an immortal pulsating universe, within which the sun and earth and supergiant red stars are comparative newcomers, the problem of initial origin remains. It merely pushes the time of Creation into the infinite past. For while theorists have adduced mathematically impeccable accounts of the fabrication of galaxies, stars, star dust, atoms, and even of the atom's components, every theory rests ultimately on the a priori assumption that *something* was already in existence—whether free neutrons, energy quanta, or simply the blank inscrutable "world stuff," the cosmic essence, of which the multifarious universe was subsequently wrought.

15

COSMOLOGISTS FOR THE MOST PART MAINTAIN silence on the question of ultimate origins, leaving that issue to the philosophers and theology. Yet only the purest empiricists among modern scientists turn their backs on the mystery that underlies physical reality. Einstein, whose philosophy of science has sometimes been criticized as materialistic, once said:

"The most beautiful and most profound emotion we can experience is the sensation of the mystical. It is the sower of all true science. He to whom this emotion is a stranger, who can no longer wonder and stand rapt in awe, is as good as dead. To know that what is impenetrable to us really exists, manifesting itself as the highest wisdom and the most radiant beauty which our dull faculties can comprehend only in their most primitive forms—this knowledge, this feeling is at the center of true religiousness."

And on another occasion he declared, "The cosmic religious experience is the strongest and noblest mainspring of scientific research." Most scientists, when referring to the mysteries of the universe, its vast forces, its origins, and its rationality and harmony, tend to avoid using the word God. Yet Einstein, who has been called an atheist, has no such inhibitions. "My religion," he says, "consists of a humble admiration of the illimitable superior spirit who reveals himself in the slight details we are able to perceive with our frail and feeble minds. That deeply emotional conviction of the presence of a superior reasoning power, which is revealed in the incomprehensible universe, forms my idea of God."

So far as science is concerned, there are at the moment two gateways which offer the promise of closer access to physical reality. One is the great new telescope which soon, from Palomar Mountain, California, will project man's vision into deeper abysses of space and time than ever were dreamed by astronomers a generation ago. Till now the extreme range of telescopic perception has essentially terminated at the faint hurrying galaxies 500 million light years away. But the two-hundred-inch reflector of Palomar will double that range, enabling man to look upon whatever lies beyond. Perhaps it will reveal only new homogeneous oceans of space and new myriads of far galaxies whose antique light has swum to earth through a billion years of terrestrial time. But it may reveal other things—variations in the density of matter or visual evidence of a cosmic curvature from which man can accurately compute the dimensions of the universe in which he so insignificantly dwells.

The other gateway to this knowledge may be opened

by the Unified Field Theory upon which Einstein has
been at work for a quarter century. Today the outer
limits of man's knowledge are defined by Relativity,
the inner limits by the Quantum Theory. Relativity
has shaped all our concepts of space, time, gravitation,
and the realities that are too remote and too vast to be
perceived. The Quantum Theory has shaped all our
concepts of the atom, the basic units of matter and
energy, and the realities that are too elusive and too
small to be perceived. Yet these two great scientific
systems rest on entirely different and unrelated theoretical foundations. The purpose of Einstein's Unified
Field Theory is to construct a bridge between them.
Believing in the harmony and uniformity of nature,
Einstein hopes to evolve a single edifice of physical
laws that will encompass both the phenomena of the
atom and the phenomena of outer space. Just as Relativity reduced gravitational force to a geometrical
peculiarity of the space-time continuum, the Unified
Field Theory will reduce electromagnetic force—the
other great universal force—to equivalent status.
"The idea that there are two structures of space independent of each other, the metric-gravitational and
the electromagnetic," Einstein observed a few years
ago, "is intolerable to the theoretical spirit." Moreover, as Relativity showed that energy has mass and
mass is congealed energy, the Unified Field Theory
will regard matter simply as a concentration of field.
From its perspective the entire universe will be revealed as an elemental field in which each star, each

atom, each wandering comet and slow-wheeling galaxy and flying electron is seen to be but a ripple or tumescence in the underlying space-time unity. And so a profound simplicity will supplant the surface complexity of nature; the distinction between gravitational and electromagnetic force, between matter and field, between electric charge and field will be forever lost; and matter, gravitation, and electromagnetic force will all thus resolve into configurations of the four-dimensional continuum which is the universe.

Completion of the Unified Field Theory will climax the long march of science towards unification of concepts. For within its framework all man's perceptions of the world and all his abstract intuitions of reality—matter, energy, force, space, time— merge finally into one. It touches the "grand aim of all science," which, as Einstein defines it, is "to cover the greatest number of empirical facts by logical deduction from the smallest possible number of hypotheses or axioms." The urge to consolidate premises, to unify concepts, to penetrate the variety and particularity of the manifest world to the undifferentiated unity that lies beyond is not only the leaven of science; it is the loftiest passion of the human intellect. The philosopher and mystic, as well as the scientist, have always sought through their various disciplines of introspection to arrive at a knowledge of the ultimate immutable essence that undergirds the mutable illusory world. More than twenty-three hundred years ago Plato declared, "The true lover of knowledge is always striving after *being*. . . . He will not rest at those multitudinous phenomena whose existence is appearance only."

*   *   *

But the irony of man's quest for reality is that as nature is stripped of its disguises, as order emerges from chaos and unity from diversity, as concepts merge and fundamental laws assume increasingly simpler form, the evolving picture becomes ever more abstract and remote from experience—far stranger indeed and less recognizable than the bone structure behind a familiar face. For where the geometry of a skull predestines the outlines of the tissue it supports, there is no likeness between the image of a tree transcribed by our senses and that propounded by wave mechanics, or between a glimpse of the starry sky on a summer night and the four-dimensional continuum that has replaced our perceptive Euclidean space.

In trying to distinguish appearance from reality and lay bare the fundamental structure of the universe, science has had to transcend the "rabble of the senses." But its highest edifices, Einstein has pointed out, have been "purchased at the price of emptiness of content." A theoretical concept is emptied of content to the very degree that it is divorced from sensory experience. For the only world man can truly know is the world created for him by his senses. If he expunges all the impressions which they translate and memory stores, nothing is left. That is what the philosopher Hegel meant by his cryptic remark: "Pure Being and Nothing are the same." A state of existence devoid of associations has no meaning. So paradoxically what the scientist and the philosopher call the world of appearance—the world of light and color, of blue skies and green leaves, of sighing wind and murmuring water, the world designed by the physiology of human sense organs—is the world in which finite man is incarcerated by his essential nature. And what the scientist and the philosopher call the world of reality—the colorless, soundless, impalpable cosmos which lies like an iceberg beneath the plane of man's perceptions—is a skeleton structure of symbols.

And the symbols change. While physicists of the last century knew, for example, that the crimson of a rose was a subjective, aesthetic sensation, they believed that "in reality" the quality they termed crimson was an oscillation of the luminiferous ether. Today it is conventional to identify crimson as a wave length. But it is equally proper to think of it as the value of the energy content of photons. Such considerations led a famous physicist to remark cynically that on Mondays, Wednesdays, and Fridays one uses the quantum theory, and on Tuesdays, Thursdays, and Saturdays the wave theory. In either case the concepts employed are abstract constructions of theory. And upon examination such concepts as gravitation, electromagnetism, energy, current, momentum, the atom, the neutron, all turn out to be theoretical substructures, inventions, metaphors which man's intellect has contrived to help him picture the true, the objective reality he apprehends beneath the surface of things. So in place of the deceitful and chaotic representations of the senses science has substituted varying systems of abstract representation. While these systems are distinguished by constantly increasing mathematical accuracy, it would be difficult today to find any scientist who imagines himself, because of his ability to discern previous errors, in a position to enunciate final truths. On the contrary, modern theorists are aware, as Newton was, that they stand on the shoulders of giants and that their particular perspective may appear as distorted to posterity as that of their predecessors seemed to them.

*   *   *

For all the promise of future revelation it is possible that certain terminal boundaries have already been reached in man's struggle to understand the manifold of nature in which he finds himself. In his descent into the microcosm he has encountered indeterminacy, duality, paradox—barriers that seem to admonish him he cannot pry too inquisitively into the heart of things without altering and vitiating the processes he seeks to observe. And in exploring the macrocosm he comes at last to a final featureless unity of space-time, mass-energy, matter-field—an ultimate, undiversified, and eternal ground beyond which there appears to be nowhere to progress. "The prison house," said Plato, "is the world of sight." Every seeming avenue of escape from this prison house that science has surveyed leads only deeper into a misty realm of symbolism and abstraction.

It may be that the extreme and insurmountable limit of scientific knowledge will be reached in the attainment of perfect isomorphic representation— that is, in a final flawless concurrence of theory and natural process, so complete that every observed phenomenon is accounted for and nothing is left out of the picture. In its approach to this goal, science has hitherto achieved its most notable pragmatic and operational triumphs. For while telling nothing of the true "nature" of things, it nevertheless succeeds in defining their relationships and depicting the events in which they are involved. "The event," Alfred North Whitehead declared, "is the unit of things real." By this he meant that however theoretical systems may change and however empty of content their symbols and concepts may be, the essential and enduring facts of science and of life are the happenings, the activities, the events. The implications of this idea can best be illustrated by contemplating a simple physical event such as the meeting of two electrons. Within the frame of modern physics one can depict this event as a collision of two elementary grains of matter or two elementary units of electrical energy, as a concourse of particles or of probability waves, or as a commingling of eddies in a four-dimensional space-time continuum. Theory does not define what the principals in this encounter actually are. Thus in a sense the electrons are not "real" but merely theoretical symbols.

On the other hand the meeting itself is "real"—the event is "real." It is as though the true objective world lies forever half-concealed beneath a translucent, plastic dome. Peering through its cloudy surface, deformed and distorted by the ever-changing perspectives of theory, man faintly espies certain apparently stable relationships and recurring events. A consistent isomorphic representation of these relationships and events is the maximal possibility of his knowledge. Beyond that point he stares into the void.

In the evolution of scientific thought, one fact has become impressively clear: there is no mystery of the physical world which does not point to a mystery beyond itself. All highroads of the intellect, all byways of theory and conjecture lead ultimately to an abyss that human ingenuity can never span. For man is enchained by the very condition of his being, his finiteness and involvement in nature. The farther he extends his horizons, the more vividly he recognizes the fact that, as the physicist Niels Bohr puts it, "we are both spectators and actors in the great drama of existence." Man is thus his own greatest mystery. He does not understand the vast veiled universe into which he has been cast for the reason that he does not understand himself. He comprehends but little of his organic processes and even less of his unique capacity to perceive the world about him, to reason and to dream. Least of all does he understand his noblest and most mysterious faculty: the ability to transcend himself and perceive himself in the act of perception.

Man's inescapable impasse is that he himself is part of the world he seeks to explore; his body and proud brain are mosaics of the same elemental particles that compose the dark, drifting dust clouds of interstellar space; he is, in the final analysis, merely an ephemeral conformation of the primordial space-time field. Standing midway between macrocosm and microcosm he finds barriers on every side and can perhaps but marvel, as St. Paul did nineteen hundred years ago, that "the world was created by the word of God so that what is seen was made out of things which do not appear."

READING LIST

This list does not include all the sources consulted in the preparation of this book; it does, however, encompass the major sources of material plus some related volumes that might be of interest and value to the reader who wishes to explore further areas of this field.

The Development of Physical Thought by Leonard B. Loeb and Arthur S. Adams. John Wiley & Sons, Inc., 1933. A survey course of modern physics. Comprehensive, beautifully written, and as enjoyable to the lay reader as valuable to the student. Specific discoveries of the last decade are, of course, not represented; otherwise a superior history of science.

"The Dust Cloud Hypothesis" by Fred L. Whipple, *Scientific American,* May 1948. It should be noted that the new *Scientific American* is no longer a gadget and hobby magazine. Under new management it now presents articles of fundamental scientific importance, brilliantly edited for intelligent laymen.

Einstein, His Life and Times by Philipp Frank. Alfred A. Knopf, 1947. The best biography of Einstein by a former colleague and eminent physicist. Detailed exposition of Einstein's scientific contributions as well as lively biographical and personal details.

The Evolution of Physics by Albert Einstein and Leopold Infeld. Simon and Schuster, 1942. The growth of ideas from early concepts to Relativity and quanta, simply and fascinatingly described. No mathematics.

Reading List

The Expanding Universe by H. P. Robertson. Reprint from *Science in Progress,* Second Series. A clear, detailed, and interesting discussion of the structure of the universe by a famed cosmologist.

Explaining the Atom by Selig Hecht. The Viking Press, 1947. Unsurpassed as an exposition of the history and the theory of the atom. Written by a famous biophysicist for the layman.

Flights from Chaos by Harlow Shapley. Whittlesey House, 1930. A fascinating survey of the universe and its component parts from atoms to galaxies by Harvard's distinguished astronomer. This book is unfortunately out of print, but should not be.

"Galaxies in Flight" by George Gamow, *Scientific American,* July, 1948. See note under *The Dust Cloud Hypothesis* above.

Meet the Atoms by O. R. Frisch. A. A. Wyn, Inc., 1947. A popular guide to modern physics. Amusingly written by a noted atomic physicist.

Mind and Nature by Hermann Weyl. University of Pennsylvania Press, 1934. The epistemological implications of modern physics set forth by a great theorist and colleague of Dr. Einstein. The philosophical standpoint from which the present volume is written is derived in great measure from this small but remarkable book, comprising five lectures delivered by Dr. Weyl at Swarthmore College in 1933.

Modern Physics by G. E. M. Jauncey. D. Van Nostrand Company, Inc., 1948. Up-to-the-minute text, including sections on nuclear physics, radioactivity, cosmic rays, and the like. Also chapters on astrophysics, Relativity, and the philosophical implications of modern science.

Mr. Tompkins in Wonderland by George Gamow. The Macmillan Company, 1947. Principles of Relativity and Quantum Theory set forth in an ingenious framework of narrative fiction.

Reading List

The Mysterious Universe by Sir James Jeans. The Macmillan Company, 1932. A classic which has awakened thousands of readers to the relationship between science and philosophy.

The Nature of the Physical World by A. S. Eddington. The Macmillan Company, 1928. Required reading for anyone interested in this field.

One Two Three . . . Infinity by George Gamow. The Viking Press, 1947. An eclectic account of facts and theories about the universe in its microscopic and macroscopic manifestations, described by a topflight theoretical physicist who is also the ablest and most entertaining expositor of science writing today.

Physics and Philosophy by Sir James Jeans. The Macmillan Company, 1943. A further discussion of the changing concepts of modern physics and their philosophical implications.

Relativity, The Special and General Theory by Albert Einstein. Henry Holt & Co., 1920. Lucid and detailed. Some mathematics, but not too difficult.

Science and the Modern World by Alfred North Whitehead. The Macmillan Company, 1925. Another classic, also required reading.

APPENDIX

I N THEORETICAL PHYSICS THERE ARE OFTEN SEV-
eral avenues of approach to a given concept. The
exposition of the principle of the increase of inertial mass on pages 55-57 follows a quickly comprehended pattern analogous to those employed by many
college physics texts. Readers with some mathematical equipment may wish to read Dr. Einstein's development of this principle as set forth in his book
Relativity, the Special and General Theory. Some essential excerpts follow, quoted with permission from
Peter Smith, Publisher.

"The most important result of a general character
to which the Special Theory of Relativity has led is
concerned with the conception of mass. Before the ad-
vent of relativity, physics recognized two conservation
laws of fundamental importance, namely the law of
the conservation of energy and the law of the conservation of mass; these two fundamental laws appeared
to be quite independent of each other. By means of the
theory of relativity they have been united into one
law. . . .

"In accordance with the theory of relativity the
kinetic energy of a material point of mass m is no

longer given by the well-known expression $m \dfrac{v^2}{2}$ but by the expression $\dfrac{mc^2}{\sqrt{1 - \dfrac{v^2}{c^2}}}$

"By means of comparatively simple considerations we are led to draw the following conclusion: A body moving with the velocity v, which absorbs an amount of energy E_0 in the form of radiation without suffering an alteration in velocity in the process, has, as a consequence, its energy increased by an amount

$$\frac{E_0}{\sqrt{1 - \dfrac{v^2}{c^2}}}$$

"In consideration of the expression given above for the kinetic energy of the body, the required energy of the body comes out to be

$$\frac{\left(m + \dfrac{E_0}{c^2}\right)C^2}{\sqrt{1 - \dfrac{v^2}{c^2}}}$$

"Thus the body has the same energy as a body of mass $\left(m + \dfrac{E_0}{c^2}\right)$ moving with the velocity v. Hence we can say: If a body takes up an amount of energy E_0, then its inertial mass increases by an amount $\dfrac{E_0}{c^2}$; the inertial mass of a body is not a constant, but varies according to the change in the energy of the body. The inertial mass of a system of bodies can even be re-garded as a measure of its energy. The law of the con-servation of the mass of a system becomes identical with the law of conservation of energy. . . ."

INDEX

Index

Index

Index

unified field, 4, 6, 107 ff.

wave, 20 ff., 110

Thermodynamics, second law of, 100

Time, 12, 39 ff., 50 ff., 61 ff., 108
  absolute, 39 ff.
  and relativity, 39 ff., 47, 50 ff.
  as form of perception, 40
  as fourth dimension, 63 ff.
  effect of gravitation on, 87 ff.
  Einstein's definition of, 39 ff., 47
  solar, 40–1, 65

Time–space continuum, 61 ff., 107 ff.

Tolman, R. C., 99

Transformation laws, 42 ff.

Ultraviolet rays, 12

Uncertainty, principle of, 26–7

Unified field theory, 4, 6, 107 ff.

Universe, and Euclidean geometry, 90 ff.
  and mathematical principles, 14, 28–9, 109 ff.
  average density of matter in, 94
  expansion of, 96 ff.
  fundamental forces, 4–5, 60, 107

  infinity of, 93 ff.
  in light of Einstein's theories, 39, 53, 60 ff., 78, 81, 89 ff., 90 ff.
  mechanical, 7 ff., 32, 34, 78
  Newtonian, 32, 34, 53, 60, 89
  origin, theories, 98 ff.
  radius of, 94

Uranium, 10, 58, 103

Velocity, of light, 35–6, 38, 39, 43–4, 46, 47, 51
  of earth through ether, 37
  of electron, 57
  of sound, 43

Wave length, of electron, 22
  of molecule, 23
  of various types of radiation, 12
  of visible light, 12

Wave mechanics, 22, 109

Wave theory, 20, 21, 33, 59, 110

Waves, electromagnetic, 13

Weight, 55

Whipple, F. L., 101

Whitehead, A. N., 112

X-rays, 13, 27